ENVIRONMENTAL REMEDIATION TECHNOLOGIES,
REGULATIONS AND SAFETY

ENVIRONMENTAL SCIENCE OF HEAVY METALS

Environmental Remediation Technologies, Regulations and Safety

Additional books and e-books in this series can be found on Nova's website under the Series tab.

ENVIRONMENTAL REMEDIATION TECHNOLOGIES,
REGULATIONS AND SAFETY

ENVIRONMENTAL SCIENCE OF HEAVY METALS

DOROTA BARTUSIK-AEBISHER
EDITOR

Copyright © 2020 by Nova Science Publishers, Inc.

All rights reserved. No part of this book may be reproduced, stored in a retrieval system or transmitted in any form or by any means: electronic, electrostatic, magnetic, tape, mechanical photocopying, recording or otherwise without the written permission of the Publisher.

We have partnered with Copyright Clearance Center to make it easy for you to obtain permissions to reuse content from this publication. Simply navigate to this publication's page on Nova's website and locate the "Get Permission" button below the title description. This button is linked directly to the title's permission page on copyright.com. Alternatively, you can visit copyright.com and search by title, ISBN, or ISSN.

For further questions about using the service on copyright.com, please contact:
Copyright Clearance Center
Phone: +1-(978) 750-8400   Fax: +1-(978) 750-4470   E-mail: info@copyright.com.

NOTICE TO THE READER

The Publisher has taken reasonable care in the preparation of this book, but makes no expressed or implied warranty of any kind and assumes no responsibility for any errors or omissions. No liability is assumed for incidental or consequential damages in connection with or arising out of information contained in this book. The Publisher shall not be liable for any special, consequential, or exemplary damages resulting, in whole or in part, from the readers' use of, or reliance upon, this material. Any parts of this book based on government reports are so indicated and copyright is claimed for those parts to the extent applicable to compilations of such works.

Independent verification should be sought for any data, advice or recommendations contained in this book. In addition, no responsibility is assumed by the Publisher for any injury and/or damage to persons or property arising from any methods, products, instructions, ideas or otherwise contained in this publication.

This publication is designed to provide accurate and authoritative information with regard to the subject matter covered herein. It is sold with the clear understanding that the Publisher is not engaged in rendering legal or any other professional services. If legal or any other expert assistance is required, the services of a competent person should be sought. FROM A DECLARATION OF PARTICIPANTS JOINTLY ADOPTED BY A COMMITTEE OF THE AMERICAN BAR ASSOCIATION AND A COMMITTEE OF PUBLISHERS.

Additional color graphics may be available in the e-book version of this book.

Library of Congress Cataloging-in-Publication Data

ISBN: 978-1-53617-831-9

Library of Congress Control Number: 2020936330

Published by Nova Science Publishers, Inc. † New York

CONTENTS

Preface		vii
Chapter 1	Microorganisms for Heavy Metal Removal *Dorota Bartusik-Aebisher, Sabina Galiniak, Tomasz Kubrak, Rafał Podgórski and David Aebisher*	1
Chapter 2	Photoactive Materials for Heavy Metal Removal *David Aebisher, Sabina Galiniak, Tomasz Kubrak, Rafał Podgórski and Dorota Bartusik-Aebisher*	53
Chapter 3	Biomass-Based Absorbents for Heavy Metal Removal *Tomasz Kubrak, Rafał Podgórski, David Aebisher, Sabina Galiniak and Dorota Bartusik-Aebisher*	87
Chapter 4	Biological Strategies for Heavy Metal Removal *Sabina Galiniak, Tomasz Kubrak, Rafał Podgórski, David Aebisher and Dorota Bartusik-Aebisher*	113

Chapter 5	Analytical Methods for the Detection and Determination of Heavy Metals in Water *Rafał Podgórski, Dominika Podgórska, David Aebisher, Sabina Galiniak, Tomasz Kubrak and Dorota Bartusik-Aebisher*	**141**

Editor Contact Information   **183**

Index   **185**

PREFACE

This book provides a current review of the problem of heavy metal removal. Microorganisms and microbial activity in environments of water and soil are presented in Chapter 1. Chapter 2 covers current knowledge about photoactive materials based on porphyrins. This book reports the manner in which plants interact with heavy metals dependent mainly on the type of contamination, species of plant as well as conditions. The book presents biological strategies for the elimination of heavy metals from polluted habitats—phytoextraction, phytostabilization, phytodegradation, phytostimulation, phytovolatilization and phytofiltration. Also included are analytical methods to determine heavy metals in water such as atomic absorption spectrometry, electrochemical methods, colorimetric and chromatographic techniques.

Chapter 1 - In this chapter the authors discuss the problem of heavy metal removal. Microorganisms and microbial activity can be visible in environment of water and soil. However, microbial activity is often disturbed by presence of the heavy metal.

Chapter 2 - The pollution of water by heavy metals is a visibly growing problem. The importance of natural water in rivers, streams, seas, and oceans is enormous, and cannot be understated. Water is the primary component of the human body, animals and plants and is the environment of many species of animals, microorganisms and plants. Water is also used in industry and is an indispensable material for human hygiene. Waterways are used for communication and transport. The development of new water purification agents based on adsorbents is required. One such material is based on silica with chemical modification with porphyrins or porphyrin derivatives. Both derivatives of porphyrin and porhyrin attached to silica are designed to be new photoactive materials and are being recognized for heavy metal removal. A primary target of purification is wastewater, which is a mixture of household waste, feces, and waste from hospitals, baths, laundries and industrial plants. A significant part of wastewater is suspended or dissolved organic compounds, mainly proteins, fats and carbohydrates. It also contains detergents, and pathogenic microorganisms that are the source of such diseases as typhus, cholera, typhoid fever, and Hein's Medina disease. Heavy metals (lead, mercury) are another important and toxic component of wastewater. These substances, when ingested by animals or humans, cause damage to the liver, blood vessels, heart, nervous system, and bones.

Chapter 3 - Continuing development of civilization and urbanization has resulted in an enormous amount of heavy metals being released into the environment. Heavy metals are highly toxic, and pose a significant threat not only to human and animal health, but also to the entire natural environment. In order to eliminate pollutants from the environment, numerous techniques have been developed using biological materials, i.e., bio-sorbents. The most important features of bio-sorbents are that they are often renewable, biodegradable and have low operating costs. Each biomass has the property of binding metal ions, however, depending on its type, the capacity to bind metals and bonding mechanisms differs. Bio-sorbents can be classified as biomass, e.g., plants (moss, leaves, trees), algae, bacteria, fungi or yeasts. Different biomass processing techniques can contribute to increasing the removal of metal ions. Research shows that various modifications of biomass significantly improve its ability to adsorb heavy metal ions.

A literature review in this chapter examines the possibilities of using and processing selected bio-sorbents for removal of heavy metal ions from polluted environments by characterizing biomass-based absorbents (e.g., aquatic biomass, terrestrial biomass, soil and mineral deposits and agricultural waste products) for heavy metal removal.

Chapter 4 - Recently, there has been a growing interest in technology called phytoremediation which includes the employment of living plant organisms for cleaning the environment. Phytoremediation technologies rely on the cultivation of species of plants that are able to grow in a contaminated environment. The manner in which the plant interacts with heavy metals depends mainly on the type of contamination, species of plant as well as conditions. Few plant species have the capability to accumulate metals in their own cells, others are able to incorporate them into their metabolic pathways. Moreover, thanks to special chemical compounds secreted by roots, these plants can cause binding of harmful substances which leads to limiting their flow into soil. These processes are the result of the natural adaptation of plants to the prevailing conditions, and these plants are treated as bioindicators of selected elements.

The chapter presents biological strategies for the elimination of heavy metals from polluted habitats including processes of phytoextraction, phytostabilization, phytodegradation, phytostimulation, phytovolatilization and phytofiltration.

Chapter 5 - Activities connected with urbanization and industrialization processes such as extensive farming, mining and chemical industry has caused release of toxic heavy metals into natural groundwater, oceans, seas, lakes, rivers, and soils. In certain amounts, several heavy metal ions are required for plant metabolism as essential micronutrients, however, they might become extremely harmful when they accumulate at high concentration in natural environments. Furthermore, they aren't biodegradable and remain in the environment for very long time. Herein are presented analytical methods used to assay levels of heavy metals in water. Heavy metal ions can be analyzed by various methods, with the choice often depending on required sensitivity and precision.

The most suitable analytical methods to determine heavy metals in water are: atomic absorption spectrometry, electrochemical methods, colorimetric and chromatographic techniques. Atomic absorption spectrometry is a very sensitive and selective technique, but it requires quite expensive equipment, use of complicated and painstaking operational procedures, and detection time is long. The advantages of electrochemical techniques include: simplicity, portability, rapid, sensitive and low-cost analysis. High performance liquid chromatography hyphenated to inductively coupled plasma mass spectrometry is currently one of the most powerful tolls for the purpose of analysis heavy metals ions in different matrices including water.

In: Environmental Science of Heavy Metals ISBN: 978-1-53617-831-9
Editor: Dorota Bartusik-Aebisher © 2020 Nova Science Publishers, Inc.

Chapter 1

MICROORGANISMS FOR HEAVY METAL REMOVAL

Dorota Bartusik-Aebisher[*], *Sabina Galiniak,*
Tomasz Kubrak, Rafał Podgórski and David Aebisher
Faculty of Medicine, University of Rzeszow, Poland

ABSTRACT

In this chapter we discuss the problem of heavy metal removal. Microorganisms and microbial activity can be visible in environment of water and soil. However, microbial activity is often disturbed by presence of the heavy metal.

Keywords: microbial activity, environment, water

The presence of heavy metals has mostly resulted from human activities and they are also naturally present in the environment. The

---

[*] Corresponding Author's Email dbartusik-aebisher@ur.edu.pl.

presence of heavy metals often coexists with other contamination events. The extent of water, soil and atmospheric contamination is large and is increasing by combustion processes in power, industrial and heat plants, road transportation, and industry. There is a need for the development of novel approaches to heavy metal (Cr, Hg, Cd, and Pb) removal from the natural environment.

At the same time, oxygen deficits occur most often in this zone, because during periods of summer or winter stratification, the oxygen supply taken in periods of circulation should be sufficient for the time when the water remains immobile (Cornu et al., 2917; Polak-Berecka et al., 2017; Parsons et al., 2016; Wang et al., 2017; Fernández et al., 2017). The self-cleaning process in stagnant water runs in completely different conditions than in flowing water, although the mechanism for removing contaminants remains the same. In deep reservoirs, the bottom zone is also devoid of sunlight, so there is no photosynthesis that could enrich water with oxygen. The inflow of pollutants causes increased biochemical oxygen demand and the formation of impurities in anaerobic conditions.

In many industrial processes, various metal ions are released. Heavy metals in most industries enter the stream of waste requiring disposal (Zheng et al., 2019; Jiang et al., 2019). The metals in sewage can also come from industry. Today, heavy metal contamination has become a serious global issue (Zheng et al., 2019; Christensen et al., 2017; Shukla et al., 2017; Shukla et al., 2017; Chen et al., 2017; Anfruns-Estrada et al., 2017).

Cadmium (Cd^{2+}) is found in wastewater from the galvanization industry, dye production industry or from production of nickel-cadmium batteries. Differences between uptake mechanisms for Pb^{2+} and Cd^{2+} in Enterobacteria has been studied (Jiang et al., 2019; Ding et al., 2018; Kolhe et al., 2018; Kheshtzar et al., 2018). Mercury mainly occurs in wastewater from acid production and from production of metallic mercury. In a study by Wu et al, the authors state that aerobic granular sludge (AGS) formed in response to a 30 g/L saline treatment exhibited the best adsorption performance (Wu et al., 2019, Cabral et al., 2019). In addition, the activity of microorganisms can be affected by surfactants which can lead to activity changes in methanogens (He et al., 2019). Lead is found in sewage from the production of batteries, dyes, brass, nickel in the galvanizing industry, paper production, refineries, steel mills and fertilizer factories. In industrial plants where sewage contains heavy metals, there should be separate technological systems to pre-treat them. Heavy metals can be removed by bio-sorbents, biomasses of fungi, seaweed, agricultural waste and remnants, yeasts, bacteria, as well as bio-sorbents containing chitosan made from shellfish coatings. Guo and coworkers showed the successful process of integration of sulfur, carbon, nitrogen and P cycles for simultaneous metabolism or removal of C, N and P (Guo et al., 2019). For example, among different methods available for remediation of Cr(VI), bioremediation is considered as one of the most sustainable methods which could effectively be utilized for controlling Cr(VI) pollution (Bhattacharya et al., 2019). Many agricultural wastes, such as fertilizers, bark or compost, contain large amounts of the substance lignin-cellulosic, which may be used in the removal of heavy metals from wastewater. For this reason, biotechnological methods are of great interest to researchers for the ability to uptake metals by microorganisms (Figueiredo et al., 2018, Zhang et al., 2019; Huang et al., 2018; Gerber et al., 2018; Li et al., 2018; Igiri et al., 2018; Rajapaksha et al., 2018; Bartolomeu et al., 2018). Bio-sorption can be described as retention of insoluble metal compounds within cellular membrabes; metal transport through cellular membranes allows for intracellular accumulation, ion-exchange adsorption or physical adsorption (Rajapaksha et al., 2018; Bartolomeu et al., 2018; Figueiredo et al., 2018; Zhang et al., 2019; Huang et al., 2018; Gerber et al., 2018; Li et al., 2018; Kheshtzar et al., 2018; Jaj et al., 2018; Zheng et al., 2017; Long et al., 2018). Heavy metals play an important role in the conductivity of solution, power generation and activity of microorganisms in bio-electrochemical systems (Zhang et al., 2018). Many factors are important in the bio-sorption process, such as pH of the solution, temperature, ionic strength or dose of bio-sorbent and initial concentration of solute.

Table 1. Presents various research topics on heavy metal removal

Author	Removal
(Fan et al., 2016) (Gaur et al., 2014) (Hechmi et al., 2014)	Fe
(Saravanan et al., 2016) (Ho et al., 2013) (Joshi et al., 2011)	Cu and Pb
(Munger et al., 2016) (Law et al., 2010) (Leles et al., 2012)	Fe and Mn
(Delforno et al., 2016) (Li et al., 2013) (Li et al., 2013) (Li et al., 2015) (Long et al., 2013) (Mulopo et al., 2013	alkylbenzene sulfonate
(Zhang et al., 2016) (Palumbo et al., 2013) (Pan et al., 2009) (Pires et al., 2011) (Sannino et al., 2010)	sulfate-reducing bacteria
(Habibul et al., 2016) (Shokri et al., 2013) (Tang et al., 2013)	Cd and Pb
(Jampasri et al., 2016) (Deary et al., 2016)	Pb
(Wagner et al., 2016)	Cu
(Li et al., 2016)	Zn
(Ma et al., 2016) (Dreannan et al., 2016) (Barboza et al., 2015)	Cd
(Wang et al., 2015)	Fe(III)
(Peng et al., 2016) (Wang et al., 2018)	chlorine-substituted phenol

Analyses of Illumina sequencing data and the relative abundance of dominant microorganisms indicate that core functional groups regulate PCP removal at genera level including Bacillus, Dethiobacter,

Desulfoporosinus and Desulfovbrio in nitrate treatments (Cheng et al., 2019). Chemical precipitation and electrochemical methods are not very effective for removal of metal from small wastewater samples.. In addition, conventional heavy metal removal technologies from wastewater generates toxic chemical deposits (Diep et al., 2018; Cheng et al., 2019; Igiri et al., 2018; Rajapaksha et al., 2018; Ding et al., 2018; Kolhe et al., 2018; Wadgaonkar et al., 2018; Madejón et al., 2018; Choińska-Pulit et al., 2018). Environmental pollution from hazardous waste materials, organic pollutants and heavy metals has adversely affected the natural ecosystem to the detriment of man. These pollutants arise from anthropogenic sources as well as natural disasters such as hurricanes and volcanic eruptions (Ojuederie et al., 2017; Nancucheo et al., 2017; Présent et al., 2017).

Graphene/NiO nanocomposites can be an innovative material to achieve complete pathogen control, along with being an economic solution for water treatment (Arshad et al., 2017). Bio-sorption on different types of biomass leads to removal, but also to metal recovery. Aspergillus aculeatus (A. aculeatus) isolated from Cd-polluted soil has been shown to increase the tolerance of turfgrasses to Cd stress (Xie et al., 2019; He et al., 2019). A study by Chen evaluated the effects of sludge lysate on anaerobic bioreduction of Cr(VI) and the role of sludge humic acid during this process (Chen et al., 2019). The biomass used in the bio-sorption process can come from waste industrial facilities available free of charge, and can also be specially raised and propagated. The abundance of electroactive bacteria, such as Acinetobacter, Pseudomonas, and Arcobacter, can be useful as bio-sorbants (Chen et al., 2019; Lin et al., 2019, Zhang et al., 2019; Simelane et al., 2019). Due to their antimicrobial properties, copper nanoparticles have been proposed to be used in agriculture for pest control. Pesticide removal is mainly performed by microorganisms (Parra et al., 2019). Fermentation processes are common in many industrial plants and the waste generated are a source of cheap bio-sorbent. The nature of the interaction between organisms and heavy metals has been defined in several articles (Ashraf et al., 2018; Diep et al., 2018). Cell walls of different species of microorganismsdiffer from each other in general composition which is the reason for the different adsorption capacity.

Asbestos fibers are highly toxic (Group 1 carcinogen) due to their high aspect ratio, durability, and the presence of iron. In nature, plants, fungi, and microorganisms release exudates, which can alter the physical and chemical properties of soil minerals including asbestos minerals (Mohanty et al., 2018; Zhang et al., 2017; Mergelsberg et al., 2017; Ma et al., 2018). Overall, one study indicated that $CuCl_2$ is toxic to anammox species, and furthermore, that EDTA attenuates $CuCl_2$ toxicity to anammox by complexing Cu^{2+} ions (Gonzalez-Estrella et al., 2017; Zhong et al., 2017; Kang et al., 2017). Also, the hydrophobicity of the cell wall, which depends on the presence of polysaccharides, proteins and lipids, affects the ability of bio-sorption. The development of adsorbents with excellent efficiency and selectivity using diverse microorganisms is ideal for treating lead pollution as described by Wang (Wang et al., 2019, Tamayo-Figueroa et al., 2019, Wu et al., 2019). Bioreactor studies of the consortia of metallic protein expressing cells immobilized on functionalized granular activated carbon revealed that 97% of copper was adsorbed from industrial effluent (Jaj et al., 2018, Zheng et al., 2017). Green zeolite can promote the translocation of V and Cd from root to shoot in Acorus calamus L., but is not conducive to Cr (Lin et al., 2019). The results showed that Fe^{2+} addition remarkably reduced the oxygen-reduction potential of both the influent and effluent water, which was beneficial to denitrification of microorganisms (Zhang et al., 2019; Simelane et al., 2019; Parra et al., 2019; Ashraf et al., 2018).

Table 2. Review of various techniques

Author	Various techniques
(Martins et al., 2017) (Bratières et al., 2012) (De Philippis et al., 2011)	antibiotics and drugs
(Mulla et al., 2016) (Klein et al, 2014) (Kuhn et al., 2014)	bacterial cultures
(Giannakis et al., 2016) (Tao et al., 2011)	study of microorganisms with micropollutants

Author	Various techniques
(Wang et al., 2013) (Wei et al., 2014) (Wiliams et al., 2013) (Wu et al., 2013) (Yan et al., 2011) (Zhang et al., 2011) (Andreazza et al., 2012) (Nevin et al., 2003) (Elias et al., 2003) (Stasinakis et al., 2003) (Lin et al., 2003) (Vainshtein et al., 2003) (Rivas et al., 2003) (Pattanapipitpaisal et al., 2002) (Mesa et al. 2002) (Tangaromsuk et al., 2002) (Ivshina et al., 2002) (Haveman et al., 2002) (Eklund et al., 2001) (Lim et al., 2002) (Watanabe et al., 2001) (Costley et al., 2001) (Lo et al., 1999) Vatsouria 2005). (Quesnel et al., 2005) (Peuke et al., 2005)	
(Hatti-Kaul et al., 2016) (Wang et al., 2013) (Wei et al., 2014)	Anaerobic microorganisms
(Mota et al., 2016) (Wiliams et al., 2013) (Wu et al., 2013)	Bio-remediation
(Kaschani et al., 2016)	iron-regulated bacterial growth in Pseudomonas aeruginosa
(Vena et al., 2016) (Fichtner et al., 2010) (Fosso-Kankeu et al., 2011)	microorganisms entrapped in biocompatible mineral matrices

Table 2. (Continued)

Author	Various techniques
(Yin et al., 2016) (Yan et al., 2011) (Zhang et al., 2011) (Andreazza et al., 2012)	biotic column vsabiotic column
(Shaw et al., 2016) (Gryzenia et al., 2009) (Hrenovic et al., 2012) (Kamika et al., 2014) (Kleinert et al., 2011) (Lira-Silva et al., 2011) (Malakahmad et al, 2011)	antibacterial studies against gram negative bacteria
(Obeid et al., 2016) (Miethke et al., 2013) (Ochoa-Herrera et al., 2011)	prokaryotic and eukaryotic organisms
(Anbu t al., 2016) (Olaniran et al., 2011) (Olguín et al., 2012)	Biomineralization
(Wu et al., 2016) (Palumbo et al., 2013) (Pan et al., 2009) (Pires et al., 2011) (Sannino et al., 2010) (hokri et al., 2013) (Tang et al., 2013)	zeolite particles.
(Mejias Carpio et al., 2016) (Orandi et al., 2013) (Pacheco et al., 2011)	Analysis of the functional groups
(Němeček et al., 2016) (Vargas-García et al., 2012) (Velimirovic et al., 2014) (Windler et al., 2013)	nano-biotechnological approaches
(Jalali et al., 2016)	Aerobic microorganisms
(Abbasian et al., 2016)	Soils contaminated with crude oil
(Mejias Carpio et al., 2016)	remediation technologies
(Liu et al., 2016)	Bio-remediation techniques
(Sun et al., 2015)	organic pollution
(Plotino et al., 2105)	planktonic microorganisms

Author	Various techniques
(Ravikumar et al., 2016) (Madden et al., 2007) (Orozco et al., 2008) (Ganesh et al., 2010) (Kim et. Al. 2011) (Pringault et al., 2010) (Velmurugan et al., 2010) (Ahammed et al., 2010)	new bio-remediation-electrokinetics
(Folgosa et al., 2015) (Stanley et al., 2003) (Ong SA, 2003). (Evdokimova et al., 2003) (Kamaludeen et al., 2003) (Vymazal et al., 2005) (Xu et al., 2005) (Amezcua-Allieri et al., 2005) (Chaudhry et al., 2005) (Ha et al., 2005) (Watanabe et al., 2004)	energy sources
(Chaturvedi et al., 2016)	heavy metal pollution
(Lkhagvajav et al., 2015)	characterization of the antimicrobial properties of nanosilver (nAg) coating on leather
(Taira et al., 2015)	Cyclic peptide of surfactin (SF)
(Mosier et al., 2015) (Jain et al., 2010) (Kuroda et al., 2010) (Bhardwaj et al., 2009) (Cotman et al., 2010) (Tapia-Rodriguez et al., 2010) (Aniszewski et al., 2010) (Martins et al., 2010)	"green" adsorption process
(Zhanng et al., 2015) (Monteiro et al., 2012) (Shelobolina et al., 2009) (Tckcrlckopoulou et al., 2013) (Alito et al., 2014) (Aziz et al., 2013) (Bai et al., 2013)	Lactococcus and Enterobacteria

Table 2. (Continued)

Author	Various techniques
(Nitzsche et al., 2015) (Baiget et al., 2013) (Bisdas et al., 2011) (Das et al., 2010) (Fan et al., 2009) (Son et al., 2006) (Lin et al., 2006) (Shrout et al., 2005) (Yang et al., 2005) (Jha et al., 2005) (Green-Ruiz et al., 2006) (Balto et al., 2005) (Nemade et al., 2009) (Bandala et al., 2009) (Zhou et al., 2009) (Muneer et al., 2009) (Franzetti et al., 2009) (Høibye et al., 2008) (Yang et al., 2009) (Simões et al., 2008) (Villegas et al., 2008) (Taşeli et al., 2008) (Sani et al., 2008) (Mertoglu et al., 2008) (Van Nooten et al., 2008) (Yoon et al., 2008) (Haferburg et al., 2007) (Vaxevanidou et al., 2008) (Srivastava et al., 2008) (Quintelas et al., 2008)	Microbial processes
(Xu et al., 2015) (González-Contreras et al., 2012) (Lizama et al., 2011) (Marková et al., 2013) (Chaput et al., 2015) (Panizza et al., 2004) (Kalin et al., 2005) (Ong et al., 2004)	biochemical processes

Author	Various techniques
(Parales et al., 2004)	
(North et al., 2004)	
(Beyenal et al., 2004)	
(Ortiz-Bernad et al., 2004)	
(Martino et al. 2004)	
(Paulo et al., 2004)	
(Zhang et al., 2003)	
(Fernandez-Sanchez et al., 2004)	
(Evdokimova et al., 2003)	
(Kamaludeen et al., 2003)	
(Nevin et al., 2003)	
(Elias et al., 2003)	
(Stasinakis et al., 2003)	
(Lin et al., 2003)	
(Vainshtein et al., 2003)	
(Rivas et al., 2003)	
(Pattanapipitpaisal et al., 2002)	
(Mesa et al., 2002)	
(Tangaromsuk et al., 2002)	
(Ivshina et al., 2002)	
(Haveman et al., 2002)	
(Eklund et al., 2001)	
(Lim et al., 2002)	
(Watanabe et al., 2001)	
(Costley et al., 2001)	
(Lo et al., 1999)	
(Mater et al., 2007)	
.(Little et al. 2007)	
(Mamais et al., 2007)	
(Yu et al., 2007)	
(Saia et al., 2007)	
(Pruden et al., 2007)	
(Congeevaram et al., 2007)	
(Ma et al., 2007)	
(Michalsen et al., 2006)	
(Botton et al., 2006)	
(Ozbelge et al., 2007)	
(Quan et al., 2006)	
(Ndjou'ou et al., 2006)	

Oxygen consumption in deep waters leads to the buildup of sulfide from sulfate reduction. Some of the microorganisms responsible for these processes also transform reactive ionic mercury to neurotoxic methylmercury (Kuss et al., 2017; Ontiveros-Valencia et al., 2017; Zhang et al., 2017; Horiike et al., 2017; Di et al., 2017; De Alencar et al., 2017). Infrared spectroscopy, studies on the effects of pH and temperature, and kinetics and isotherm modelling are carried out to evaluate the bio-sorption. Infrared spectroscopy shows that the primary bio-sorption sites are carboxylate groups (Luk et al., 2017; Dai et al., 2017; Fan et al., 2017; Moreno-Sánchez et al., 2017). The dispersion of granules in an upflow anaerobic sludge blanket reactor represents a critical technical issue in methanolic wastewater treatment (Zhen, et al. 2107). The reduction of Cr(VI) by microorganisms has been found to be insignificant, indicating the adsorption/co-precipitation of Cr by iron oxides on iron surfaces was responsible for overall Cr(VI) removal (Yin et al., 2017; Zhang et al., 2017; Giovanella et al., 2017; Ontañon et al., 2017). The ability of Klebsiella species 3S1 to form silver chloride nanoparticles with interesting potential applications has also been discussed (Muñoz et al., 2017; Ayangbenro et al., 2017; Lukić et al. 2017). Table 2 reviews various techniques used for eco-friendly bio-sorbant technology (Wang et al., 2015; Gerbino et al., 2015; Kiran et al., 2016; Hassan et al., 2015; Saha et al., 2016; Ma et al., 2015; Sivrioğlu et al., 2015; Mirazimi et al., 2015; Amor et al., 2015; Yuan et al., 2015; Zuo et al., 2015, Fang et al., 2015, Christoffels et al., 2014, Joutey et al., 2015; Wang et al., 2015, (Arjomandzadegan et al., 2014, Wang et al., 2014). Nitrogen removal efficiency and N_2O production during the process of coupling catalytic iron and biological denitrification for low C/N ratio wastewater were studied (Mackie et al., 2014, Ye et al., 2015, Michailides et al., 2015, Rhee 2014; Fan et al., 2014). The mechanism of adsorption was determined by studies of carboxylic, hydroxyl, and amino organic compounds (Sowmya et al., 2014; Tartanson et al., 2014, Jiang et al., 2014, Wang et al., 2014). The study of cadmium (Cd) and pyrene (Pyr) were also performed (Javanbakht et al., 2014; Ye et al., 2014, Hassana et al., 2014; Navarro et al., 2013; Yang et al.,2014; Karmous et al., 2014, Kamika et al., 2014; Alito et al.,

2014). The results shown in one study indicate that wastewater treatment could be impacted by Ag and AgNPs in the short term, but the amount of treatment disruption will depend on the magnitude of influent Ag (Gaur et al., 2014). Heavy metals such as Pb, As, Hg and Cd are serious threats to public health since they are implicated in cellular oxidative stress and cell death (Kuhn et al., 2014). Microorganisms can change the redox states of iron and sulfur resulting in iron and sulfur compounds with low solubility leading to precipitation (Klein et al., 2014; Velimirovic et al., 2014; Shokri et al., 2013, Wei et al., 2014). Biodegradation of both bio-accessible and associated pyrene fractions was enhanced by celery rhizodeposition in pyrene spiked soils (Wang et al., 2013). The promoting effects of the two additives on recovering microbial activity and removing excessive biomass were also observed in an article by Wu et al. (Wu et al., 2013, Li et al., 2013). This work might offer valuable implications for the optimization and practical application of ZVI-anaerobic sludge processes for treatment of azo dyes or other recalcitrant pollutants (Tang et al., 2013; Mulopo et al., 2013). In reactor B, sulphate removal efficiency was accompanied by an accumulation of COD as hydrogen (H_2) provided by Fe(0) was utilized for sulphate reduction. Furthermore, these results showed the potential of Fe(0) to enhance the participation of microorganisms in sulphate reduction (Palumbo et al., 2013). Such reactors could maintain water quality for closed-cycle bio-refineries, leading to reduced water consumption, and a more sustainable biofuel (Li et al., 2015). The problems and future perspectives of this technology have been discussed (Wiliams et al., 2013; Hechmi et al., 2014; Ho et al., 2013; Li et al., 2013; Long et al., 2013). Results have suggested that the indigenous bacterial strain LY6 can be used for bio-remediation (Chua et al., 1999; Obara et al., 1999, Bach et al., 1999). A comparison of silver-coated and uncoated central venous catheters regarding bacterial colonization was performed to assess the relative contribution of catheter hub and skin colonization to catheter tip colonization (Deming et al., 1998; Ekstrand et al., 1998).

According to the conclusions of independent evaluations from different state health agencies, the release of mercury from dental amalgam does not present any non-acceptable risk to the general population (Gupta et al., 1998; Matzanke et al., 1997). The ultimate storage compound is an E. coli-type bacterioferritin, in which over 90% of cellular iron is located (Edlund et al., 1996). Mercury resistance together with multiple antimicrobial resistance was observed and reported (Rubin et al., 1995; Tai et al., 1995). In the subunit of Salmonella typhimurium alkyl hydroperoxide reductase and polypeptides of other microorganisms associated with oxidation reduction, activity was observed (Molloy et al., 1990). Iron from catalyzing undesirable oxidative reactions, as well as making it unavailable for growth of microorganisms that survive the killing process (Macaskie et al., 1990; Schweisfurth et al., 1989; Maki et al., 1988; Tabatowski et al., 1988; Arnold et al., 1982; Limsuwan et al., 1981; Tengerdy et al., 1981; Zelepukha et al., 1975; Thompson et al., 1971).

ACKNOWLEDGMENT

Dorota Bartusik-Aebisher acknowledges support from the National Center of Science NCN (New drug delivery systems-MRI study, Grant OPUS-13 number 2017/25/B/ST4/02481).

REFERENCES

Abbasian F, Palanisami T, Megharaj M, Naidu R, Lockington R, Ramadass K (2016). Microbial diversity and hydrocarbon degrading gene capacity of a crude oil field soil as determined by metagenomics analysis. *Biotechnology Progress.* 32(3):638-48.

Ahammed MM, Meera V (2010) Metal oxide/hydroxide-coated dual-media filter for simultaneous removal of bacteria and heavy metals from natural waters. *Journal of Hazardous Materials.* 181(1-3):788-93.

Alito CL, Gunsch CK (2014) Assessing the effects of silver nanoparticles on biological nutrient removal in bench-scale activated sludge sequencing batch reactors. *International Journal of Environmental Science and Technology.* 48(2):970-6.

Amezcua-Allieri MA, Lead JR, Rodríguez-Vázquez R (2005) Changes of chromium behavior in soil during phenanthrene removal by Penicillium frequentans. *Biometals.* 18(1):23-9.

Amor C, Lucas MS, García J, Dominguez JR, De Heredia JB, Peres JA (2015) Combined treatment of olive mill wastewater by Fenton's reagent and anaerobic biological process. *Journal of Environmental Science and Health, Part A. Environ Eng.* 50(2):161-8.

Anbu P, Kang CH, Shin YJ, So JS (2016) Formations of calcium carbonate minerals by bacteria and its multiple applications. *Springerplus.* 5:250.

Andreazza R, Okeke BC, Pieniz S, Camargo FA(2012). Characterization of copper-resistant rhizosphere bacteria from Avena sativa and Plantago lanceolata for copper bioreduction and biosorption. *Biological Trace Element Research.* 146(1):107-15.

Anfruns-Estrada E, Bruguera-Casamada C, Salvadó H, Brillas E, Sirés I, Araujo RM (2017) Inactivation of microbiota from urban wastewater by single and sequential electrocoagulation and electro-Fenton treatments. *Water Research.* 126:450-459.

Aniszewski E, Peixoto RS, Mota FF, Leite SG, Rosado AS (2010) Bioemulsifier production byMicrobacterium SP. strains isolated from mangrove and their application to remove cadmiun and zinc from hazardous industrial residue. *Braz Journal of Microbiology and Biotechnology.* 41(1):235-45.

Arjomandzadegan M, Rafiee P, Moraveji MK, Tayeboon M (2014). Efficacy evaluation and kinetic study of biosorption of nickel and zinc by bacteria isolated from stressed conditions in a bubble column. *Asian Pac Journal of Tropical Medicine.* 7S1:S194-8.

Arnold RR, Russell JE, Champion WJ, Brewer M, Gauthier JJ (1982) Bactericidal activity of human lactoferrin: differentiation from the stasis of iron deprivation. *Infection and immunity.* 35(3):792-9.

Arshad A, Iqbal J, Mansoor Q (2017) NiO-nanoflakes grafted graphene: an excellent photocatalyst and a novel nanomaterial for achieving complete pathogen control. *Nanoscale.* 9(42):16321-16328.

Ashraf S., Naveed M, Afzal M, Ashraf S, Rehman K, Hussain A, Zahir ZA (2018) Bioremediation of tannery effluent by Cr- and salt-tolerant bacterial strains. *Environmental Monitoring and Assessment.* 190(12):716.

Ayangbenro AS, Babalola OO (2017) A New Strategy for Heavy Metal Polluted Environments: A Review of Microbial Biosorbents. *International Journal of Environmental Research and Public Health.* 14(1).

Aziz HA, Othman OM, Abu Amr SS (2013). The performance of Electro-Fenton oxidation in the removal of coliform bacteria from landfill leachate. *Waste Management.* 33(2):396-400.

Bach A, Eberhardt H, Frick A, Schmidt H, Böttiger BW, Martin E (1999) Efficacy of silver-coating central venous catheters in reducing bacterial colonization. *Critical Care Medicine.* 27(3):515-21.

Bai Y, Liu R, Liang J, Qu J (2013) Integrated metagenomic and physiochemical analyses to evaluate the potential role of microbes in the sand filter of a drinking water treatment system. *PLoS One.* 8(4):e61011.

Baiget M, Constantí M, López MT, Medina F (2013) Uranium removal from a contaminated effluent using a combined microbial and nanoparticle system. *New Biotechnology.* 30(6):788-92.

Balto H, Al-Nazhan S, Al-Mansour K, Al-Otaibi M, Siddiqu Y (2005) Microbial leakage of Cavit, IRM, and Temp Bond in post-prepared root canals using two methods of gutta-percha removal: an in vitro study. *The Journal of Contemporary Dental Practice.* 6(3):53-61.

Bandala ER, Miranda J, Beltran, M, Vaca M, López R., Torres LG (2009) Wastewater disinfection and organic matter removal using ferrate (VI) oxidation. *Journal of Water and Health.* 7(3):507-13.

Barboza NR, Amorim SS, Santos PA, Reis FD, Cordeiro MM., Guerra-Sá R, Leão VA (2015) Indirect Manganese Removal by Stenotrophomonas sp. and Lysinibacillus sp. Isolated from Brazilian Mine Water. *BioMed Research International* 2015:925972.

Barry AN, Shinde U, Lutsenko S (2010) Structural organization of human Cu-transporting ATPases: learning from building blocks. *Journal of Biological Inorganic Chemistry.* (1):47-59.

Bartolomeu M, Neves MGPMS, Faustino MAF, Almeida A (2018) Wastewater chemical contaminants: remediation by advanced oxidation processes. *Photochemical and Photobiological Sciences* 17(11):1573-1598.

Bartolomeu M, Neves MGPMS, Faustino MAF, Almeida A (2018) Wastewater chemical contaminants: remediation by advanced oxidation processes. *Photochemical and Photobiological Sciences.* 17(11):1573-1598.

Beyenal H, Lewandowski Z (2004) Dynamics of lead immobilization in sulfate reducing biofilms. *Water Research.* 38(11):2726-36.

Bhardwaj SB, Mehta M, Gauba K (2009) Nanotechnology: role in dental biofilms. *Indian Journal of Dental Research.* 20(4):511-3.

Bhattacharya A, Gupta A, Kaur A, Malik D (2019) Alleviation of hexavalent chromium by using microorganisms: insight into the strategies and complications. *Water Science and Technology.* 79 (3):411-424.

Bisdas T, Wilhelmi M, Haverich A, Teebken OE (2011) Cryopreserved arterial homografts vs silver-coated Dacron grafts for abdominal aortic infections with intraoperative evidence of microorganisms. *Journal of Vascular Surgery.* 53(5):1274-1281.e4.

Botton S, Parsons JR (2006) Degradation of btex compounds under iron-reducing conditions in contaminated aquifer microcosms. *Environmental Toxicology and Chemistry.* 25(10):2630-8.

Bratières K, Schang C, Deletić A, McCarthy DT (2012) Performance of enviss™ stormwater filters: results of a laboratory trial. *Water Science and Technology.* 66(4):719-27.

Cabral L, Noronha MF, De Sousa STP, Lacerda-Júnior GV, Richter L, Fostier AH, Andreote FD, Hess M, Oliveira VM (2019) The metagenomic landscape of xenobiotics biodegradation in mangrove sediments. *Ecotoxicol and Environ Safety.* 179:232-240.

Chaput DL, Hansel CM, Burgos WD, Santelli CM (2015) Profiling microbial communities in manganese remediation systems treating coal mine drainage. *Applied Environmental Microbiology.*81(6): 2189-98.

ChaturvediAD, Pal D, Penta S, Kumar A (2016) Ecotoxic heavy metals transformation by bacteria and fungi in aquatic ecosystem. *World Journal of Microbiology and Biotechnology.* 31(10):1595-603.

Chaudhry Q, Blom-Zandstra M, Gupta S, Joner EJ (2005) Utilising the synergy between plants and rhizosphere microorganisms to enhance breakdown of organic pollutants in the environment. *Environmental Science and Pollution Research (international).* 12(1):34-48.

Chen H, Jin R, Liu G, Tian T, Gu C, Zhou J, Xing D (2019) Effects of sludge lysate for Cr(VI) bioreduction and analysis of bioaugmentation mechanism of sludge humic acid. *Environmental Science and Pollution Research.* 26(5):5065-5075.

Chen X, Song D, Xu J, Sun G, Xu M (2017) Microbial depassivation of Fe(0) for contaminant removal under semi-aerobic conditions. *Applied Microbiology and Biotechnology.* 101(23-24):8595-8605.

Cheng J, Xue L, Zhu M, Feng J, Shen-Tu J, Xu J, Brookes PC, Tang C, He Y (2019) Nitrate supply and sulfate-reducing suppression facilitate the removal of pentachlorophenol in a flooded mangrove soil. *Environmental Pollution.* 244:792-800.

Choińska-Pulit A, Sobolczyk-Bednarek J, Łaba W (2018) Optimization of copper, lead and cadmium biosorption onto newly isolated bacterium using a Box-Behnken design. *Ecotoxicology and Environmental Safety.* 149:275-283.

Christensen E, Nilsen V, Håkonsen T, Heistad A, Gantzer C, Robertson LJ, Myrmel M (2017) Removal of model viruses, E. coli and Cryptosporidium oocysts from surface water by zirconium and chitosan coagulants *Journal of Water and Health.* 15(5):695-705.

Christoffels E, Mertens FM, Kistemann T, Schreiber C (2014) Retention of pharmaceutical residues and microorganisms at the Altendorf retention soil filter *Water Science Technology.* 70(9):1503-9.

Chua H, Yu PH, Sin SN, Cheung MW (1999) Sub-lethal effects of heavy metals on activated sludge microorganisms. *Chemosphere.* 39(15):2681-92.

Congeevaram S, Dhanarani S, Park J, Dexilin M, Thamaraiselvi K (2007) Biosorption of chromium and nickel by heavy metal resistant fungal and bacterial isolates *Journal of Hazardous Materials.* 146(1-2):270-7.

Cornu JY, Huguenot D, Jézéquel K, Lollier, M, Lebeau, T (2017) Bioremediation of copper-contaminated soils by bacteria *World Journal of Microbiology and Biotechnology* 33(2):26.

Costley SC, Wallis FM (2001) Bioremediation of heavy metals in a synthetic wastewater using a rotating biological contactor *Water Research* 35(15):3715-23.

Costley SC, Wallis FM (2001) Bioremediation of heavy metals in a synthetic wastewater using a rotating biological contactor *Water Research* 35(15):3715-23.

Cotman M, Gotvajn AZ (2010) Comparison of different physico-chemical methods for the removal of toxicants from landfill leachate *Journal of Hazardous Materials* 178(1-3):298-305.

Dai S, Li Y, Zhou T, Zhao Y (2017) Reclamation of heavy metals from contaminated soil using organic acid liquid generated from food waste: removal of Cd Cu and Zn and soil fertility improvement *Environmental Science and Pollution Research* 24(18):15260-15269.

Das D, Das N, Mathew L (2010) Kinetics equilibrium and thermodynamic studies on biosorption of Ag(I) from aqueous solution by macrofungus Pleurotus platypus *Journal of Hazardous Materials* 184(1-3):765-74.

De Alencar FLS, Navoni JA, do Amaral VS (2017) The use of bacterial bioremediation of metals in aquatic environments in the twenty-first century: a systematic review *Environmental Science and Pollution Research* 24(20):16545-16559.

De Philippis R, Colica G, Micheletti E (2011) Exopolysaccharide-producing cyanobacteria in heavy metal removal from water: molecular basis and practical applicability of the biosorption process *Applied Microbiology and Biotechnology* 92(4):697-708.

Deary ME, Ekumankama CC, Cummings SP 2016) Development of a novel kinetic model for the analysis of PAH biodegradation in the presence of lead and cadmium co-contaminants *Journal of Hazardous Materials* 307:240-52.

Delforno TP, Okada DY, Faria CV, Varesche MBA (2016) Evaluation of anionic surfactant removal in anaerobic reactor with Fe(III) supplementation *Journal of Environmental Management* 183(Pt 3):687-693.

Deming JW (1998) Deep ocean environmental biotechnology *Current Opinion in Biotechnology* 9(3):283-7.

Di J, Wang M (2017) Experimental study on treating agate dyeing wastewater with sulfate-reducing bacteria strengthening peanut shells and scrap iron *Water Science Technology* 76(3-4):939-952.

Diep P, Mahadevan R, Yakunin A F (2018) Heavy Metal Removal by Bioaccumulation Using Genetically Engineered Microorganisms *Frontiers in Bioengineering and Biotechnology* 6:157.

Ding L, Tan WF, Xie SB, Mumford K, Lv JW, Wang HQ, Fang Q, Zhang XW, Wu XY, Li M (2018) Uranium adsorption and subsequent re-oxidation under aerobic conditions by Leifsonia sp - Coated biochar as green trapping agent *Environmental Pollution* 242(Pt A):778-787.

Drennan DM, Almstrand R, Lee I, Landkamer L, Figueroa L, Sharp JO (2016) Organoheterotrophic Bacterial Abundance Associates with Zinc Removal in Lignocellulose-Based Sulfate-Reducing Systems *Environmental Science and Technology* 50(1):378-87.

Edlund C, Björkman L, Ekstrand J, Sandborgh-Englund G, Nord CE (1996) Resistance of the normal human microflora to mercury and antimicrobials after exposure to mercury from dental amalgam fillings *Clinical Infectious Diseases* 22(6):944-50.

Eklund H, Uhlin U, Färnegårdh M, Logan DT, Nordlund P (2001) Structure and function of the radical enzyme ribonucleotide reductase *Progress in Biophysics and Molecular Biology* 77(3):177-268.

Eklund H, Uhlin U, Färnegårdh M, Logan DT, Nordlund P (2001) Structure and function of the radical enzyme ribonucleotide reductase *Progress in Biophysics and Molecular Biology* 77(3):177-268.

Ekstrand J, Björkman L, Edlund C, Sandborgh-Englund G (1998) Toxicological aspects on the release and systemic uptake of mercury from dental amalgam *European Journal of Oral Sciences* 106(2 Pt 2):678-86.

Elias DA, Krumholz LR, Wong D, Long PE Suflita JM (2003) Characterization of microbial activities and U reduction in a shallow aquifer contaminated by uranium mill tailings *Microbial Ecology* 46(1):83-91.

Elias DA, Krumholz LR, Wong D, Long PE, Suflita JM (2003) Characterization of microbial activities and U reduction in a shallow aquifer contaminated by uranium mill tailings *Microbial Ecology* 46(1):83-91.

Evdokimova GA, Mozgova NP (2003) Restoration of properties of cultivated soils polluted by copper and nickel *Journal of Environmental Monitoring* 5(4):667-70.

Fan ZX, Zhao X, Wang JL (2009) Denitrification using radiation-pretreated wheat straw as solid carbon source *Huan Jing Ke Xue* 30(4):1090-4.

Fan JH, Hao RX, Li M, Zhu XX, Wan JJ (2016) Phosphorus Removal Mechanism of Sulfur/Sponge Iron Composite Fillers Based on Denitrification. *Huan Jing Ke Xue* 4275-4281.

Fan J, Onal Okyay T, Frigi Rodrigues D (2014) The synergism of temperature pH and growth phases on heavy metal biosorption by two environmental isolates *Journal of Hazardous Materials* 279:236-43.

Fang J, Su B, Sun P, Lou J, Han J (2015) Long-term effect of low concentration Cr(VI) on P removal in granule-based enhanced biological phosphorus removal (EBPR) system *Chemosphere* 121:76-83.

Fernández PM, Cruz EL, Viñarta SC, Castellanos de Figueroa LI (2017) Optimization of Culture Conditions for Growth Associated with Cr(VI) Removal by Wickerhamomyces anomalus M10 *Bulletin of Environmental Contamination and Toxicology* 98(3):400-406.

Fernandez-Sanchez JM, Sawvel EJ, Alvarez PJ (2004) Effect of Fe0 quantity on the efficiency of integrated microbial-Fe0 treatment processes *Chemosphere* 54(7):823-9.

Fichtner J, Güresir E, Seifert V, Raabe A (2010) Efficacy of silver-bearing external ventricular drainage catheters: a retrospective analysis *Journal of Neurosurgical Sciences* 112(4):840-6.

Figueiredo N, Serralheiro ML, Canário J, Duarte A, Hintelmann H, Carvalho C (2018) Evidence of Mercury Methylation and Demethylation by the Estuarine Microbial Communities Obtained in Stable Hg Isotope Studies International Journal of Environmental Research and Public Health 15(10) Folgosa F, Tavares P, Pereira AS (2015) Iron management and production of electricity by microorganisms *Appl Microbiol Biotechnol* 99(20):8329-36.

Fosso-Kankeu E, Mulaba-Bafubiandi AF, Mamba BB, Barnard TG (2011) Prediction of metal-adsorption behaviour in the remediation of water contamination using indigenous microorganisms *Journal of Environmental Management* 92(10):2786-93.

Franzetti A, Caredda P, Ruggeri C, La Colla P, Tamburini E, Papacchini M, Bestetti G (2009) Potential applications of surface active compounds by Gordonia sp strain BS29 in soil remediation technologies *Chemosphere* 75(6):801-7.

Ganesh R, Smeraldi J, Hosseini T, Khatib L, Olson BH, Rosso D (2010) Evaluation of nanocopper removal and toxicity in municipal wastewaters *Environmental Science and Technology* 44(20):7808-13.

Gaur N, Flora G, Yadav M, Tiwari A (2014) A review with recent *advancements on bioremediation-based abolition of heavy metals Environmental Science: Processes and Impacts* 16(2):180-93.

Gerber U, Hübner R, Rossberg A, Krawczyk-Bärsch E, Merroun ML (2018) Metabolism-dependent bioaccumulation of uranium by Rhodosporidium toruloides isolated from the flooding water of a former uranium mine *PLoS One* 13(8):e0201903.

Gerbino E, Carasi P, Araujo-Andrade C, Tymczyszyn EE, Gómez-Zavaglia A (2015) Role of S-layer proteins in the biosorption capacity of lead by Lactobacillus kefir *World Journal of Microbiology and Biotechnology* 31(4):583-92.

Giannakis S, Voumard M, Grandjean D, Magnet A De, Alencastro LF, Pulgarin C (2016) Micropollutant degradation bacterial inactivation and regrowth risk in wastewater effluents: Influence of the secondary (pre)treatment on the efficiency of Advanced Oxidation Processes *Water Research* 102:505-515.

Giovanella P, Cabral L, Costa AP, de Oliveira Camargo FA, Gianello C Bento FM (2017) Metal resistance mechanisms in Gram-negative bacteria and their potential to remove Hg in the presence of other metals *Ecotoxicology and Environmental Safety* 140:162-169.

González-Contreras P, Weijma J, Buisman CJ (2012) Continuous bioscorodite crystallization in CSTRs for arsenic removal and disposal *Water Research* 46(18):5883-92.

Gonzalez-Estrella J, Li G, Neely SE, Puyol D Sierra-Alvarez R, Field JA (2017) Elemental copper nanoparticle toxicity to anaerobic ammonium oxidation and the influence of ethylene diamine-tetra acetic acid (EDTA) on copper toxicity *Chemosphere* 184:730-737.

Green-Ruiz C (2006) Mercury(II) removal from aqueous solutions by nonviable Bacillus sp from a tropical estuary *Bioresource Technology* 97(15):1907-11.

Gryzenia J, Cassidy D, Hampton D (2009) Production and accumulation of surfactants during the chemical oxidation of PAH in soil Chemosphere 77(4):540-5.

Guo G, Ekama GA, Wang Y, Dai J, Biswal BK, Chen G, Wu D (2019) Advances in sulfur conversion-associated enhanced biological phosphorus removal in sulfate-rich wastewater treatment: A review *Bioresource Technology*.

Gupta G, Keegan B (1998) Bioaccumulation and biosorption of lead by poultry litter microorganisms *Poultry Science Journal* 77(3):400-4.

Gupta A, Dutta A, Sarkar J, Panigrahi MK, Sar P (2018) Low-Abundance Members of the Firmicutes Facilitate Bioremediation of Soil Impacted by Highly Acidic Mine Drainage From the Malanjkhand Copper Project India *Frontiers in Microbiology* 9:2882.

Ha KY, Chung YG, Ryoo SJ (2005) Adherence and biofilm formation of Staphylococcus epidermidis and Mycobacterium tuberculosis on various spinal implants *Spine* (Phila Pa 1976) 30(1):38-43.

Habibul N, Hu Y, Sheng GP (2016) Microbial fuel cell driving electrokinetic remediation of toxic metal contaminated soils. *Journal of Hazardous Materials* 318:9-14.

Haferburg G, Merten D, Büchel G, Kothe E (2007) Biosorption of metal and salt tolerant microbial isolates from a former uranium mining area Their impact on changes in rare earth element patterns in acid mine drainage *Journal of Basic Microbiology* 47(6):474-84.

Hassan Z, Sultana M, Van Breukelen BM, Khan S, Röling WF (2015) Diverse arsenic- and iron-cycling microbial communities in arsenic-contaminated aquifers used for drinking water in Bangladesh *FEMS Microbial Ecology* 91(4).

Hassana SR, Zwaina HM, Zamana NQ, Dahlanb I (2014) Recycled paper mill effluent treatment in a modified anaerobic baffled reactor: start-up and steady-state performance *Environmental Technology* 35(1-4):294-9.

Hatti-Kaul R, Mattiasson B (2016) Anaerobes in Industrial- and Environmental Biotechnology Advanced Biochemical Engineering *Biotechnology* 156:1-33.

Haveman SA, Pedersen K (2002) Microbially mediated redox processes in natural analogues for radioactive waste *Journal of Contaminant Hydrology* 55(1-2):161-74.

He Q, Xu P, Zhang C, Zeng G, Liu Z, Wang D, Tang W, Dong H, Tan X, Duan A (2019) Influence of surfactants on anaerobic digestion of waste activated sludge: acid and methane production and pollution removal *Critical Reviews in Biotechnology* 39(5):746-757.

He Z, Li Z, Zhang Q, Wei Z, Duo J, Pan X (2019) Simultaneous remediation of As(III) and dibutyl phthalate (DBP) in soil by a manganese-oxidizing bacterium and its mechanisms *Chemosphere* 220:837-844.

Hechmi N, Aissa NB, Abdenaceur H, Jedidi N (2014) Evaluating the phytoremediation potential of Phragmites australis grown in pentachlorophenol and cadmium co-contaminated soils *Environmental Science and Pollution Research* 21(2):1304-13.

Ho KL, Lin WC, Chung YC, Chen YP, Tseng CP (2013) Elimination of high concentration hydrogen sulfide and biogas purification by chemical-biological process *Chemosphere* 92(10):1396-401.

Høibye L Clauson-Kaas J Wenzel H Larsen HF Jacobsen BN Dalgaard O (2008) Sustainability assessment of advanced wastewater treatment technologies *Water Science and Technology* 58(5):963-8.

Horiike T, Dotsuta Y, Nakano Y, Ochiai A, Utsunomiya S, Ohnuki T, Yamashita M, (2017) Removal of Soluble Strontium via Incorporation into Biogenic Carbonate Minerals by Halophilic Bacterium Bacillus sp Strain TK2d in a Highly Saline Solution *Applied and Environmental Microbiology* 83(20).

Hrenovic J, Milenkovic J, Daneu N, Kepcija RM, Rajic N (2012) Antimicrobial activity of metal oxide nanoparticles supported onto natural clinoptilolite *Chemosphere* 88(9):1103-7.

Huang H, Zhao Y, Xu Z, Ding Y, Zhang W, Wu L (2018) Biosorption characteristics of a highly Mn(II)-resistant Ralstonia pickettii strain isolated from Mn ore *PLoS One* 13(8):e0203285.

Huang J, Cao C, Liu J, Yan C, Xiao J (2019) The response of nitrogen removal and related bacteria within constructed wetlands after long-term treating wastewater containing environmental concentrations of silver nanoparticles *Science of the Total Environment* 667:522-531.

Igiri BE, Okoduwa SIR, Idoko GO, Akabuogu EP, Adeyi AO, Ejiogu IK (2018) Toxicity and Bioremediation of Heavy Metals Contaminated Ecosystem from Tannery Wastewater: A Review *Journal of Toxicology* 2018:2568038.

Ivshina IB, Peshkur TA, Korobov VP (2002) The efficient accumulation of cesium ions by Rhodococcus cells *Mikrobiologiia* 71(3):418-23.

Jamiagsagssr PSL, Manrkkg NA (2018) Biomimetic strategies to design metallic proteins for detoxification of hazardous heavy metal *Journal of Hazardous Materials* 358:92-100.

Jain R, Pathak A, Sreekrishnan TR, Dastidar MG (2010) Autoheated thermophilic aerobic sludge digestion and metal bioleaching in a two-stage reactor system *Journal of Environmental Sciences (China)* 22(2):230-6.

Jalali J, Magdich S, Jarboui R, Loungou M, Ammar E (2016) Phosphogypsum biotransformation by aerobic bacterial flora and isolated Trichoderma asperellum from Tunisian storage piles *Journal of Hazardous Materials* 308:362-73.

Jampasri K, Pokethitiyook P, Kruatrachue M, Ounjai P, Kumsopa A (2016) Phytoremediation of fuel oil and lead co-contaminated soil by Chromolaena odorata in association with Micrococcus luteus *International Journal of Phytoremediation* 18(10):994-1001.

Javanbakht V Alavi S A Zilouei H (2014) Mechanisms of heavy metal removal using microorganisms as biosorbent *Water Science and Technology* 69(9):1775-87.

Jha D, Bose P (2005) Development of an advanced biological treatment system applied to the removal of nitrogen and phosphorus using the sludge ceramics *Chemosphere* 61(7):1020-31.

Jiang R, Qi J, Wang W, Zheng H, Li X (2014) Accumulation and fraction distribution of Ni(II) in activated sludge treating Ni-laden wastewater *Environmental Science and Pollution Research* 21(18):10744-50.

Joshi PK, Swarup A, Maheshwari S, Kumar R, Singh N (2011) Bioremediation of heavy metals in liquid media through fungi isolated from contaminated sources *Indian Journal of Microbiology* 51(4): 482-7.

Joutey NT, Sayel H, Bahafid WE, Ghachtouli N (2015) Mechanisms of hexavalent chromium resistance and removal by microorganisms *Reviews of Environmental Contamination and Toxicology* 233:45-69.

Kalin M, Wheeler WN, Meinrath G (2005) The removal of uranium from mining waste water using algal/microbial biomass *Journal of Environmental Radioactivity* 78(2):151-77.

Kamaludeen SP, Megharaj M, Juhasz AL, Sethunathan N, Naidu R (2003) Chromium-microorganism interactions in soils: remediation implications *Reviews of Environmental Contamination and Toxicology* 178:93-164.

Kamika I, Momba MN (2014) icrobial diversity of Emalahleni mine water in South Africa and tolerance ability of the predominant organism to vanadium and nickel *PLoS One* 9(1):e86189.

Kang S M Jang S C Heo N S Oh S Y Cho H J Rethinasabapathy M Vilian ATE Han Y K Roh C Huh Y S (2017) Cesium-induced inhibition of bacterial growth of Pseudomonas aeruginosa PAO1 and their possible potential applications for bioremediation of wastewater *Journal of Hazardous Materials* 338:323-333.

Karmous I, Chaoui A, Jaouani K, Sheehan D, El Ferjani E, Scoccianti V, Crinelli R (2014) Role of the ubiquitin-proteasome pathway and some peptidases during seed germination and copper stress in bean cotyledons *Plant Physiology and Biochemistry* 76:77-85.

Kaschani F, Wei Q, Dingemans J, Van der Hoorn RA, Cornelis P Kaiser M (2016) Capture of endogenously biotinylated proteins from Pseudomonas aeruginosa displays unexpected downregulation of LiuD upon iron nutrition *Bioorganic and Medicinal Chemistry* 24(15):3330-5.

Kheshtzar I, Ghorbani M, Gatabi MP, Lashkenari MS (2018) Facile synthesis of smartaminosilane modified- SnO2/porous silica nanocomposite for high efficiency removal of lead ions and bacterial inactivation *Journal of Hazardous Materials* 359:19-30.

Kim HJ, Du W, Ismagilov RF (2011) Complex function by design using spatially pre-structured synthetic microbial communities: degradation of pentachlorophenol in the presence of Hg(ii) *Integrative Biology (Camb)* 3(2):126-33.

Kiran GS, Ninawe AS, Lipton AN, Pandian V, Selvin J (2016) Rhamnolipid biosurfactants: evolutionary implications applications and future prospects from untapped marine resource *Critical Reviews in Biotechnology* 36(3):399-415.

Klein R, Tischler JS, Mühling M, Schlömann M (2014) Bioremediation of mine water *Advances in Biochemical Engineering/Biotechnology* 141:109-72.

Kleinert S, Muehe EM, Posth NR, Dippon U, Daus B, Kappler A (2011) Biogenic Fe(III) minerals lower the efficiency of iron-mineral-based commercial filter systems for arsenic removal *Environmental Science and Technology* 45(17):7533-41.

Kolhe N, Zinjarde S, Acharya C (2018) Responses exhibited by various microbial groups relevant to uranium exposure *Biotechnology Advances* 36(7):1828-1846.

Kuhn KM, Maurice PA, Neubauer E, Hofmann T, von der Kammer F (2014) Accessibility of humic-associated Fe to a microbial siderophore: implications for bioavailability *International Journal of Environmental Science and Technology* 48(2):1015-22.

Kumar S, Dutta V (2019) Constructed wetland microcosms as sustainable technology for domestic wastewater treatment: an overview *Environmental Science and Pollution Research* 26(12):11662-11673.

Kuroda K, Ueda M (2010) Engineering of microorganisms towards recovery of rare metal ions *Applied Microbiology and Biotechnology* 87(1):53-60.

Kuss J, Cordes F, Mohrholz V, Nausch G, Naumann M, Krüger S, Schulz-Bull D E (2017) The Impact of the Major Baltic Inflow of December 2014 on the Mercury Species Distribution in the Baltic Sea *International Journal of Environmental Science and Technology* 51(20):11692-11700.

Law GT, Geissler A, Lloyd JR, Livens FR, Boothman C, Begg JD, Denecke MA, Rothe J Dardenne K, Burke IT, Charnock JM, Morris K (2010) Geomicrobiological redox cycling of the transuranic element neptunium *Environmental Science and Technology* 44(23):8924-9.

Leles DM, Lemos DA, Filho UC, Romanielo LL, de Resende MM, Cardoso VL (2012) Evaluation of the bioremoval of Cr(VI) and TOC in biofilters under continuous operation using response surface methodology *Biodegradation* 23(3):441-54.

Li C, Xu Y, Jiang W, Dong X, Wang D, Liu B (2013) Effect of NaCl on the heavy metal tolerance and bioaccumulation of Zygosaccharomyces rouxii and Saccharomyces cerevisiae *Bioresource Technology* 143:46-52.

Li PS, Tao HC (2015) Cell surface engineering of microorganisms towards adsorption of heavy metals *Critical Reviews in Microbiology* 41(2):140-9.

Li WW, Zhang Y, Zhao JB, Yang YL, Zeng RJ, Liu HQ, Feng YJ (2013) Synergetic decolorization of reactive blue 13 by zero-valent iron and anaerobic sludge *Bioresource Technology* 149:38-43.

Li C, Yang X,, Xu Y Li L, Wang Y (2018) Cadmium detoxification induced by salt stress improves cadmium tolerance of multi-stress-tolerant Pichia kudriavzevii *Environmental Pollution* 242(Pt A):845-854.

Li C, Yu J,, Wang D Li L, Yang X Ma H, Xu Y (2016) Efficient removal of zinc by multi-stress-tolerant yeast Pichia kudriavzevii A16 *Bioresource Technology* 206:43-49.

Lim PE, Ong SA Seng CE (2002) Simultaneous adsorption and biodegradation processes in sequencing batch reactor (SBR) for treating copper and cadmium-containing wastewater *Water Research* 36(3):667-75.

Limsuwan T, Lovell RT (1981) Intestinal synthesis and absorption of vitamin B-12 in channel catfish *Journal of Nutrition* 111(12):2125-32.

Lin Q, Wang Z, Ma S, Chen Y (2006) Evaluation of dissipation mechanisms by Lolium perenne L and Raphanus sativus for pentachlorophenol (PCP) in copper co-contaminated soil *Science of the Total Environment* 368(2-3):814-22.

Lin YM, Yang XF, Liu Y (2003) Kinetic responses of activated sludge microorganisms to individual and joint copper and zinc *J Environ Sci Health A Tox Hazard Subst Environ Eng* 38(2):353-60.

Lin YM, Yang XF, Liu Y (2003) Kinetic responses of activated sludge microorganisms to individual and joint copper and zinc *J Environ Sci Health Journal of Environmental Science and Health* Part A Toxic/hazardous substances and environmental engineering 38(2): 353-60.

Lin H, Liu J, Dong Y, He Y (2019) The effect of substrates on the removal of low-level vanadium chromium and cadmium from polluted river water by ecological floating beds *Ecotoxicology and Environmental Safety* 169:856-862.

Lira-Silva E, Ramírez-Lima, IS Olín-Sandoval, V García-García, JD García-Contreras R Moreno-Sánchez R Jasso-Chávez R (2011) Removal accumulation and resistance to chromium in heterotrophic Euglena gracilis *Journal of Hazardous Materials* 193:216-24.

Little B, Lee J, Ray R (2007) A review of 'green' strategies to prevent or mitigate microbiologically influenced corrosion *Biofouling* 23(1-2):87-97.

Liu J, Liu W, Wang F, Kerr P, Wu Y (2016) Redox zones stratification and the microbial community characteristics in a periphyton bioreactor *Bioresource Technology* 204:114-121.

Lizama A K Fletcher TD Sun G (2011) Removal processes for arsenic in constructed wetlands *Chemosphere* 84(8):1032-43.

Lkhagvajav N, Koizhaiganova M, Yasa I, Çelik E, Sari Ö (2015) Characterization and antimicrobial performance of nano silver coatings on leather materials *Brazilian Journal of Microbiology* 46(1):41-8.

Lo W, Chua H, Lam KH, Bi SP (1999) A comparative investigation on the biosorption of lead by filamentous fungal biomass *Chemosphere* 39(15):2723-36.

Long D, Tang X, Cai K, Chen G, Shen C, Shi J, Chen L, Chen Y (2013) Cr(VI) resistance and removal by indigenous bacteria isolated from chromium-contaminated soil *Journal of Microbiology and Biotechnology* 23(8):1123-32.

Long D, Tang X, Cai K, Chen G, Shen C, Shi J, Chen L, Chen Y (2013) Cr(VI) resistance and removal by indigenous bacteria isolated from chromium-contaminated soil *Journal of Microbiology and Biotechnology Reports* 23(8):1123-32.

Long B, Ye B, Liu Q, Zhang S, Ye J, Zou L, Shi J (2018) Characterization of Penicillium oxalicum SL2 isolated from indoor air and its application to the removal of hexavalent chromium *PLoS One* 13(1):e0191484.

Lu M, Zhang Z, Qiao W, Wei X, Guan Y, Ma Q, Guan Y (2010) Remediation of petroleum-contaminated soil after composting by sequential treatment with Fenton-like oxidation and biodegradation *Bioresource Technology* 101(7):2106-13.

Luk CHJ, Yip J, Yuen CWM, Pang SK, Lam KH, Kan CW (2017) Biosorption Performance of Encapsulated Candida krusei for the removal of Copper(II) *Scientific Reports* 7(1):2159.

Lukić B, Huguenot, D Panico A, van Hullebusch ED, Esposito G (2017) Influence of activated sewage sludge amendment on PAH removal efficiency from a naturally contaminated soil: application of the landfarming treatment *Environmental Technology* 38(23):2988-2998.

Ma H, Wang B, Wang Y (2007) Application of molybdenum and phosphate modified kaolin in electrochemical treatment of paper mill wastewater *Journal of Hazardous Materials* 145(3):417-23.

Ma L, Wang F, Yu Y, Liu J, Wu Y (2018) Cu removal and response mechanisms of periphytic biofilms in a tubular bioreactor *Bioresource Technology* 2018 Jan;248(Pt B):61-67.

Ma N, Li C, Dong X, Wang D, Xu Y (2015) Different effects of sodium chloride preincubation on cadmium tolerance of Pichia kudriavzevii and Saccharomyces cerevisiae *Journal of Basic Microbiology* 55(8):1002-12.

Ma X, Guo N, Ren S, Wang S, Wang Y (2019) Response of antibiotic resistance to the co-existence of chloramphenicol and copper during bio-electrochemical treatment of antibiotic-containing wastewater *Environment International* 126:127-133.

Macaskie LE (1990) An immobilized cell bioprocess for the removal of heavy metals from aqueous flows *Journal of Chemical Technology and Biotechnology* 49(4):357-79.

Mackie K A Schmidt H P Müller T Kandeler E (2014) Cover crops influence soil microorganisms and phytoextraction of copper from a moderately contaminated vineyard *Science of the Total Environment* 1;500-501:34-43.

Madden AS, Smith AC, Balkwill DL, Fagan LA, Phelps TJ (2007) Microbial uranium immobilization independent of nitrate reduction *Environ Microbiol* 9(9):2321-30.

Madejón P, Domínguez MT, Madejón E, Cabrera F, Marañón T, Murillo JM (2018) Soil-plant relationships and contamination by trace elements: A review of twenty years of experimentation and monitoring after the Aznalcóllar (SW Spain) mine accident *Science of the Total Environment* 625:50-63.

Maki DG, Cobb L, Garman JK, Shapiro JM, Ringer M, Helgerson RB (1988) An attachable silver-impregnated cuff for prevention of infection with central venous catheters: a prospective randomized multicenter trial *The American Journal of Medicine* 85(3):307-14.

Malakahmad A, Hasani A, Eisakhani M, Isa MH (2011) Sequencing Batch Reactor (SBR) for the removal of Hg^{2+} and Cd^{2+} from synthetic petrochemical factory wastewater *Journal of Hazardous Materials* 191(1-3):118-25.

Mamais D, Noutsopoulos C, Andreadakis A, Droubogianni J, Georgakopoulos A, Tsepapadakis E Mariolos J (2007) *Environmental Technology* 28(2):129-36.

Marková Z, Šišková KM, Filip J, Čuda J, Kolář M, Šafářová K, Medřík I, Zbořil R (2013) Air stable magnetic bimetallic Fe-Ag nanoparticles for advanced antimicrobial treatment and phosphorus removal *Environmental Science and Technology* 47(10):5285-93.

Martino E, Cerminara S, Prandi L, Fubini B, Perotto S (2004) Physical and biochemical interactions of soil fungi with asbestos fibers *Environmental Toxicology and Chemistry* 23(4):938-44.

Martins M, Faleiro ML, Chaves S, Tenreiro R, Santos E, Costa MC (2010) Anaerobic bio-removal of uranium (VI) and chromium (VI): comparison of microbial community structure *Journal of Hazardous Materials* 176(1-3):1065-72.

Martins M, Mourato C, Sanches S, Noronha JP, Crespo MTB, Pereira IAC (2017) Biogenic platinum and palladium nanoparticles as new catalysts for the removal of pharmaceutical compounds *Water Research* 108:160-168.

Mater L, Rosa EV, Berto J, Corrêa AX, Schwingel PR, Radetski CM (2007) A simple methodology to evaluate influence of $H2O2$ and $Fe(2+)$ concentrations on the mineralization and biodegradability of organic compounds in water and soil contaminated with crude petroleum *Journal of Hazardous Materials* 149(2):379-86.

Matzanke BF, Böhnke R, Möllmann U, Reissbrodt R, Schünemann V, Trautwein AX (1997) Iron uptake and intracellular metal transfer in mycobacteria mediated by xenosiderophores *Biometals* 10(3):193-203.

Mejias Carpio IE, Franco DC, Zanoli Sato MI, Sakata S Pellizari VH, Seckler Ferreira Filho S, Frigi Rodrigues D (2016) Biostimulation of metal-resistant microbial consortium to remove zinc from contaminated environments *Science of The Total Environment* 550:670-675.

Mergelsberg M, Willistein M, Meyer H, Stär HJ, Bechtel DF, Pierik AJ, Boll M (2017) Phthaloyl-coenzyme A decarboxylase from Thauera chlorobenzoica: the prenylated flavin- K+ - and Fe2+ -dependent key enzyme of anaerobic phthalate degradation *Environmental Microbiology* 19(9):3734-3744.

Mertoglu B, Semerci N, Guler N, Calli B, Cecen F, Saatci AM (2008) Monitoring of population shifts in an enriched nitrifying system under gradually increased cadmium loading *Journal of Hazardous Materials* 160(2-3):495-501.

Mesa MM, Macías M Cantero D (2002) Mathematical model of the oxidation of ferrous iron by a biofilm of Thiobacillus ferrooxidans *Biotechnology Progress* 18(4):679-85.

Michailides MK, Tekerlekopoulou AG, Akratos CS, Coles S, Pavlou S, Vayenas DV (2015) Molasses as an efficient low-cost carbon source for biological Cr(VI) removal *J Hazard Mater* 281:95-105.

Rhee YJ, Hillier S, Pendlowski H, Gadd GM (2014) Fungal transformation of metallic lead to pyromorphite in liquid medium *Chemosphere* 113:17-21.

Michalsen MM, Goodman BA, Kelly SD, Kemner KM, McKinley JP, Stucki JW, Istok JD (2006) Uranium and technetium bio-immobilization in intermediate-scale physical models of an in situ bio-barrier *Environmental Science and Technology* 40(22):7048-53.

Miethke M, (2013) Molecular strategies of microbial iron assimilation: from high-affinity complexes to cofactor assembly systems *Metallomics* 5(1):15-28.

Mirazimi SM, Abbasalipour Z, Rashchi F (2015) Vanadium removal from LD converter slag using bacteria and fungi *Journal of Environmental Management* 153:144-51.

Mohanty SK Gonneau C Salamatipour A Pietrofesa RA Casper B Christofidou-Solomidou M Willenbring JK (2018) Siderophore-mediated iron removal from chrysotile: Implications for asbestos toxicity reduction and bioremediation *Journal of Hazardous Materials* 341:290-296.

Molloy AL, Winterbourn CC (1990) Release of iron from phagocytosed Escherichia coli and uptake by neutrophil lactoferrin *Blood* 75(4): 984-9.

Monteiro CM, Brandão TR, Castro PM, Malcata FX (2012) Modelling growth of and removal of Zn and Hg by a wild microalgal consortium *Applied Microbiology and Biotechnology* 94(1):91-100.

Moreno-Sánchez R, Rodríguez-Enríquez S, Jasso-Chávez R, Saavedra E, García-García JD (2017) Biochemistry and Physiology of Heavy Metal Resistance and Accumulation in Euglena *Advances in Experimental Medicine and Biology* 979:91-121.

Mosier AP, Behnke J, Jin ET, Cady NC (2015) Microbial biofilms for the removal of Cu^{2+} from CMP wastewater *Journal of Environmental Management* 160:67-72.

Mota R, Rossi F, Andrenelli L, Pereira SB, De Philippis R, Tamagnini P (2016) Released polysaccharides (RPS) from Cyanothece sp CCY 0110 as biosorbent for heavy metals bioremediation: interactions between metals and RPS binding sites *Applied Microbiology and Biotechnology* 100(17):7765-75.

Mulla SI, Sun Q, Hu A, Wang Y, Ashfaq M, Eqani SA, Yu CP (2016) Evaluation of Sulfadiazine Degradation in Three Newly Isolated Pure Bacterial Cultures *PLoS One* 11(10):e0165013.

Mulopo J, Schaefer L (2013) Effect of the addition of zero valent iron (Fe(0)) on the batch biological sulphate reduction using grass cellulose as carbon source *Applied Biochemistry and Biotechnology* 171(8):2020-9.

Muneer B, Rehman A, Shakoori FR, Shakoori AR (2009) Evaluation of consortia of microorganisms for efficient removal of hexavalent chromium from industrial wastewater *Bulletin of Environmental Contamination and Toxicology* 82(5):597-600.

Munger ZW, Carey CC, Gerling AB, Hamre KD, Doubek JP, Klepatzki SD, McClure RP, Schreiber ME (2016) Effectiveness of hypolimnetic oxygenation for preventing accumulation of Fe and Mn in a drinking water reservoir *Water Research* 106:1-14.

Muñoz AJ, Espínola F, Ruiz E (2017) Biosorption of Ag(I) from aqueous solutions by Klebsiella sp 3S1 *Journal of Hazardous Materials* 329:166-177.

Nancucheo I, Bitencourt JAP, Sahoo PK, Alves JO, Siqueira JO, Oliveira G (2017) Recent Developments for Remediating Acidic Mine Waters Using Sulfidogenic Bacteria *BioMed Research International* 2017:7256582.

Navarro CA, Von Bernath D, Jerez CA (2013) Heavy metal resistance strategies of acidophilic bacteria and their acquisition: importance for biomining and bioremediation *Biological Research* 46(4):363-71.

Ndjou'ou AC, Cassidy D (2006) Surfactant production accompanying the modified Fenton oxidation of hydrocarbons in soil *Chemosphere* 65(9):1610-5.

Nemade PD, Kadam AM, Shankar HS (2009) Wastewater renovation using constructed soil filter (CSF): a novel approach *Journal of Hazardous Materials* 170(2-3):657-65.

Němeček J, Pokorný P, Lhotský O, Knytl V, Najmanová P, Steinová J, Černík M, Filipová A Filip J, Cajthaml T (2016) Combined nano-biotechnology for in-situ remediation of mixed contamination of groundwater by hexavalent chromium and chlorinated solvents *Science of The Total Environment* 563-564:822-34.

Nevin KP, Finneran KT, Lovley DR (2003) Microorganisms associated with uranium bioremediation in a high-salinity subsurface sediment *Applied Environmental Microbiology* 69(6):3672-5.

Nitzsche KS, Weigold P, Lösekann-Behrens T, Kappler A, Behrens S (2015) Microbial community composition of a household sand filter used for arsenic iron and manganese removal from groundwater in Vietnam *Chemosphere* 138:47-59.

North NN, Dollhopf SL, Petrie L, Istok JD, Balkwill DL, Kostka JE (2004) Change in bacterial community structure during in situ biostimulation of subsurface sediment cocontaminated with uranium and nitrate *Applied and Environmental Microbiology* 70(8):4911-20.

Obara T, Sawaya T, Hokari K, Umehara Y, Mizukami M, Tomita F (1999) Removal of cadmium from scallop hepatopancreas by microbial processes *Bioscience Biotechnology and Biochemistry* 63(3):500-5.

Obeid MH, Oertel J, Solioz M, Fahmy K (2016) Mechanism of Attenuation of Uranyl Toxicity by Glutathione in Lactococcus lactis *Applied and Environmental Microbiology* 82(12):3563-3571.

Ochoa-Herrera V, León G, Banihani Q, Field JA, Sierra-Alvarez R (2011) Toxicity of copper(II) ions to microorganisms in biological wastewater treatment systems *Science of the Total Environment* 412-413:380-5.

Ojuederie OB, Babalola OO (2017) Microbial and Plant-Assisted Bioremediation of Heavy Metal Polluted Environments: A Review *Int J Environ Res Public Health* 14(12).

Olaniran AO, Balgobind A, Pillay B (2011) Quantitative assessment of the toxic effects of heavy metals on 12-dichloroethane biodegradation in co-contaminated soil under aerobic condition *Chemosphere* 85(5): 839-47.

Olguín EJ, Sánchez-Galván G (2012) Heavy metal removal in phytofiltration and phycoremediation: the need to differentiate between bioadsorption and bioaccumulation *New Biotechnology* 30(1):3-8.

Ong SA, Lim PE, Seng CE (2003) Effects of adsorbents and copper(II) on activated sludge microorganisms and sequencing batch reactor treatment process *Journal of Hazardous Materials* 103(3):263-77.

Ong SA, Toorisaka E, Hirata M, Hano T (2004) Effects of nickel(II) addition on the activity of activated sludge microorganisms and activated sludge process *Journal of Hazardous Materials* 113(1-3):111-21.

Ontañon OM, González PS, Barros GG, Agostini E (2017) Improvement of simultaneous Cr(VI) and phenol removal by an immobilised bacterial consortium and characterisation of biodegradation products *New Biotechnology* 37(Pt B):172-179.

Ontiveros-Valencia A, Zhou C, Ilhan ZE, De Saint Cyr LC, Krajmalnik-Brown R, Rittmann BE (2017) Total electron acceptor loading and composition affect hexavalent uranium reduction and microbial community structure in a membrane biofilm reactor *Water Res* 125:341-349.

Orandi S, Lewis DM (2013) Biosorption of heavy metals in a photo-rotating biological contactor--a batch process study *Applied Microbiology and Biotechnology* 97(11):5113-23.

Orozco AM, Contreras EM, Zaritzky NE (2008) Modelling Cr(VI) removal by a combined carbon-activated sludge system *Journal of Hazardous Materials* 150(1):46-52.

Ortiz-Bernad I, Anderson RT, Vrionis HA, Lovley DR (2004) Vanadium respiration by Geobacter metallireducens: novel strategy for in situ removal of vanadium from groundwater *Applied and Environmental Microbiology* 70(5):3091-5.

Ozbelge TA, Ozbelge HO, Altinten P (2007) Effect of acclimatization of microorganisms to heavy metals on the performance of activated sludge process *Journal of Hazardous Materials* 142(1-2):332-9.

Pacheco PH, Gil RA, Cerutti SE, Smichowski P, Martinez LD (2011) Biosorption: a new rise for elemental solid phase extraction methods *Talanta* 85(5):2290-300.

Palumbo AJ, Daughney CJ, Slade AH, Glover CN (2013) Influence of pH and natural organic matter on zinc biosorption in a model lignocellulosic biofuel biorefinery effluent *Bioresource Technology* 146:169-175.

Pan R, Cao L, Zhang R (2009) Combined effects of Cu Cd Pb and Zn on the growth and uptake of consortium of Cu-resistant Penicillium sp A1 and Cd-resistant Fusarium sp A19 *Journal of Hazardous Materials* 171(1-3):761-6.

Panizza M, Cerisola G (2004) Electrochemical oxidation as a final treatment of synthetic tannery wastewater *International Journal of Environmental Science and Technology* 38(20):5470-5.

Parales RE, Haddock JD (2004) Biocatalytic degradation of pollutants *Current Opinion in Biotechnology* 15(4):374-9.

Parra B, Tortella GR, Cuozzo S Martínez M (2019) Negative effect of copper nanoparticles on the conjugation frequency of conjugative catabolic plasmids *Ecotoxicology and Environmental Safety* 169:662-668.

Parsons D, Meredith K, Rowlands VJ, Short D, Metcalf DG, Bowler PG (2016) Enhanced Performance and Mode of Action of a Novel Antibiofilm Hydrofiber® Wound Dressing *Biomed Research International* 2016:7616471.

Pattanapipitpaisal P, Mabbett AN, Finlay JA, Beswick AJ, Paterson-Beedle M, Essa A, Wright J, Tolley MR, Badar U, Ahmed N, Hobman JL, Brown NL, Macaskie LE (2002) Reduction of Cr(VI) and bioaccumulation of chromium by gram positive and gram negative microorganisms not previously exposed to Cr-stress *Journal Environmental Technology* 23(7):731-45.

Paulo PL, Jiang B, Cysneiros D, Stams AJ, Lettinga G (2004) Effect of cobalt on the anaerobic thermophilic conversion of methanol *Biotechnology and Bioengineering* 85(4):434-41.

Peng J, Li J, Shi H, Wang Z, Gao S (2016) Oxidation of disinfectants with Cl-substituted structure by a Fenton-like system $Cu(2+)/H_2O_2$ and analysis on their structure-reactivity relationship *Environmental Science and Pollution Research* 23(2):1898-904.

Peuke AD, Rennenberg H (2005) Phytoremediation with transgenic trees *Z Naturforsch* C 60(3-4):199-207.

Pires C, Marques AP, Guerreiro A,, Magan N Castro PM (2011) Removal of heavy metals using different polymer matrixes as support for bacterial immobilisation *Journal of Hazardous Materials* 191(1-3):277-86.

Plotino G, Ahmed HM, Grande NM, Cohen S, Bukiet F (2015) Current Assessment of Reciprocation in Endodontic Preparation: A

Comprehensive Review--Part II: Properties and Effectiveness *Journal of Endodontics* 41(12):1939-50.

Polak-Berecka M, Boguta P, Cieśla J, Bieganowski A, Skrzypek T, Czernecki T, Waśko A (2017) Studies on the removal of Cd ions by gastrointestinal lactobacilli *Applied Microbiology and Biotechnology* 101(8):3415-3425.

Présent RM, Rotureau E, Billard P, Pagnout C, Sohm B, Flayac J, Gley R, Pinheiro JP, Duval JFL (2017) Impact of intracellular metallothionein on metal biouptake and partitioning dynamics at bacterial interfaces *Physical Chemistry Chemical Physics* 19(43):29114-29124.

Pringault O, Viret H, Duran R (2010) Influence of microorganisms on the removal of nickel in tropical marine sediments (New Caledonia) *Marine Pollution Bulletin* 61(7-12):530-41.

Pruden A, Messner N, Pereyra L, Hanson RE, Hiibel SR, Reardon KF (2007) The effect of inoculum on the performance of sulfate-reducing columns treating heavy metal contaminated water *Water Research* 41(4):904-14.

Quan X, Tan H, Zhao Y, Hu Y (2006) Detoxification of chromium slag by chromate resistant bacteria *Journal of Hazardous Materials* 137(2):836-41.

Quesnel D, Nakhla G (2005) Optimization of the aerobic biological treatment of thermophilically treated refractory wastewater *Journal of Hazardous Materials* 125(1-3):221-30.

Quintelas C, Fernandes B, Castro J, Figueiredo H, Tavares T (2008) Biosorption of Cr(VI) by three different bacterial species supported on granular activated carbon: a comparative study *Journal of Hazardous Materials* 153(1-2):799-809.

Rajapaksha PP, Power A, Chandra S, Chapman J (2018) Graphene electrospun membranes and granular activated carbon for eliminating heavy metals pesticides and bacteria in water and wastewater treatment processes *Analyst* 143(23):5629-5645.

Ravikumar K V Kumar D Rajeshwari A Madhu G M Mrudula P Chandrasekaran N Mukherjee A (2016) A comparative study with biologically and chemically synthesized nZVI: applications in Cr (VI) removal and ecotoxicity assessment using indigenous microorganisms from chromium-contaminated site *Environ Environmental Science and Pollution Research* 23(3):2613-27.

Rivas FJ, Beltrán FJ, Gimeno O, Alvarez P (2003) Optimisation of Fenton's reagent usage as a pre-treatment for fermentation brines *Journal of Hazardous Materials* 96(2-3):277-90.

Rubin RB, Barton AL, Banner BF, Bonkovsky HL (1995) Iron and chronic viral hepatitis: emerging evidence for an important interaction *Digestive Diseases* 13(4):223-38.

Saha M, Sarkar S, Sarkar B, Sharma BK, Bhattacharjee S, Tribedi P (2016) Microbial siderophores and their potential applications: a review *Environmental Science and Pollution Research* 23(5):3984-99.

Saia FT, Damianovic MH, Cattony EB, Brucha G, Foresti E, Vazoller RF (2007) Anaerobic biodegradation of pentachlorophenol in a fixed-film reactor inoculated with polluted sediment from Santos-São Vicente Estuary Brazil *Applied Microbiology and Biotechnology* 75(3):665-72.

Sani RK, Peyton BM, Dohnalkova A (2008) Comparison of uranium(VI) removal by Shewanella oneidensis MR-1 in flow and batch reactors *Water Research* 42(12):2993-3002.

Sannino F, Pirozzi D, Aronne A, Fanelli E, Spaccini R, Yousuf A, Pernice P (2010) Remediation of waters contaminated with MCPA by the yeasts Lipomyces starkeyi entrapped in a sol-gel zirconia matrix *Environmental Science and Technology* 44(24):9476-81.

Saravanan R, Ravikumar L (2016) Cellulose bearing Schiff base and carboxylic acid chelating groups: a low cost and green adsorbent for heavy metal ion removal from aqueous solution *Water Science Technology* 74(8):1780-1792.

Schweisfurth R (1989) Microbiologic studies of in situ removal of iron and manganese *Schriftenr Ver Wasser Boden Lufthyg* 80:167-86.

Shaw R, Sharma R, Tiwari S, Tiwari SK (2016) Surface Engineered Zeolite: An Active Interface for Rapid Adsorption and Degradation of Toxic Contaminants in Water ACS *ACS Applied Materials and Interfaces* 8(19):12520-7.

Shelobolina ES, Konishi H, Xu H, Roden EE (2009) U(VI) sequestration in hydroxyapatite produced by microbial glycerol 3-phosphate metabolism *Applied Environmental Microbiology* 75(18):5773-8.

Shokri M, Jodat A, Modirshahla N, Behnajady MA (2013) Photocatalytic degradation of chloramphenicol in an aqueous suspension of silver-doped TiO2 nanoparticles *Journal Environmental Technology* 34(9-12):1161-6.

Shrout JD, Larese-Casanova P, Scherer MM, Alvarez PJ (2005) Sustained and complete hexahydro-135-trinitro-135-triazine (RDX) degradation in zero-valent iron simulated barriers under different microbial conditions *Environmental Technology* 26(10):1115-26.

Shukla A, Parmar P, Saraf M (2017) Radiation radionuclides and bacteria: An in-perspective review *Journal of Environmental Radioactivity* 180:27-35.

Simelane S, Dlamini LN (2019) An investigation of the fate and behaviour of a mixture of WO3 and TiO2 nanoparticles in a wastewater treatment plant *Journal of Environmental Sciences* (China) 76:37-47.

Simões M, Simoes LC, Pereira MO, Vieira MJ (2008) Antagonism between Bacillus cereus and Pseudomonas fluorescens in planktonic systems and in biofilms *Biofouling* 24(5):339-49.

Sivrioğlu Ö, Yonar T (2015) Determination of the acute toxicities of physicochemical pretreatment and advanced oxidation processes applied to dairy effluents on activated sludge *Journal of Dairy Science* 98(4):2337-44.

Solá SMZ, Lovaisa N, Dávila Costa JS, Benimeli CS, Polti MA, Alvarez A (2019) Multi-resistant plant growth-promoting actinobacteria and plant root exudates influence Cr(VI) and lindane dissipation *Chemosphere* 222:679-687.

Son A, Lee J, Chiu PC, Kim BJ, Cha DK (2006) Microbial reduction of perchlorate with zero-valent iron *Water Research* 40(10):2027-2032.

Sowmya M, Rejula MP, Rejith PG, Mohan M, Karuppiah M, Hatha AA (2014) Heavy metal tolerant halophilic bacteria from Vembanad Lake as possible source for bioremediation of lead and cadmium *Journal of Environmental Biology* 35(4):655-60.

Srivastava NK, Majumder CB (2008) Novel biofiltration methods for the treatment of heavy metals from industrial wastewater *Journal of Hazardous Materials* 151(1):1-8.

Stanley LC, Ogden KL (2003) Biosorption of copper (II) from chemical mechanical planarization wastewaters *Journal of Environmental Management* 69(3):289- 97.

Stanley LC, Ogden KL (2003) Biosorption of copper (II) from chemical mechanical planarization wastewaters *Journal of Environmental Management* 69(3):289-97.

Stasinakis AS, Thomaidis NS, Mamais D, Papanikolaou EC,, Tsakon A Lekkas TD (2003) Effects of chromium (VI) addition on the activated sludge process *Water Research* 37(9):2140-8.

Sun L, Li Y, Li A (2015) Treatment of Actual Chemical Wastewater by a Heterogeneous Fenton Process Using Natural Pyrite *International Journal of Environmental Research and Public Health* 12(11): 13762-78.

Tabatowski K, Roggli VL, Fulkerson WJ, Langley RL, Benning T, Johnston WW (1988) Giant cell interstitial pneumonia in a hard-metal worker Cytologic histologic and analytical electron microscopic investigation *Acta Cytologica* 32(2):240-6.

Tai SS, Zhu YY (1995) Cloning of a Corynebacterium diphtheriae iron-repressible gene that shares sequence homology with the AhpC subunit of alkyl hydroperoxide reductase of Salmonella typhimurium *Journal of Bacteriology* 177(12):3512-7.

Taira T, Yanagisawa S, Nagano T, Zhu Y, Kuroiwa T, Koumura N, Kitamoto D, Imura T (2015) Selective encapsulation of cesium ions using the cyclic peptide moiety of surfactin: Highly efficient removal based on an aqueous giant micellar system *Colloids Colloids and Surfaces B: Biointerfaces* 134:59-64.

Tamayo-Figueroa DP, Castillo E, Brandão PFB (2019) Metal and metalloid immobilization by microbiologically induced carbonates precipitation *World Journal of Microbiology and Biotechnology* 35(4):58.

Tandukar M, Huber SJ, Onodera T, Pavlostathis SG (2009) Biological chromium(VI) reduction in the cathode of a microbial fuel cell *Environmental Science and Technology* 43(21):8159-65.

Tang J Wang ZX Xu XH (2013) Removal of Cr(VI) by iron filings with microorganisms to recover iron reactivity] *Huan Jing Ke Xue* 34(7):2650-7.

Tangaromsuk J, Pokethitiyook P, Kruatrachue M, Upatham ES (2002) Cadmium biosorption by Sphingomonas paucimobilis biomass *Bioresource Technology* 85(1):103-5.

Tao HC, Zhang LJ, Gao ZY, Wu WM (2011) Copper reduction in a pilot-scale membrane-free bioelectrochemical reactor *Bioresource Technology* 102(22):10334-9.

Tapia-Rodriguez A, Luna-Velasco A, Field JA,, Sierra-Alvarez R (2010) Anaerobic bioremediation of hexavalent uranium in groundwater by reductive precipitation with methanogenic granular sludge *Water Research* 44(7):2153-62.

Tartanson MA, Soussan L, Rivallin M, Chis C, Penaranda D, Lapergue R, Calmels P, Faur C (2014) A new silver based composite material for SPA water disinfection *Water Research* 63:135-46.

Taşeli BK, Gökçay CF, Gürol A (2008) Influence of nickel (II) and chromium (VI) on the laboratory scale rotating biological contactor *Journal of Industrial Microbiology and Biotechnology* 35(9):1033-9.

Tekerlekopoulou AG, Tsiflikiotou M, Akritidou L, Viennas A, Tsiamis G, Pavlou S, Bourtzis K, Vayenas DV (2013) Modelling of biological Cr(VI) removal in draw-fill reactors using microorganisms in suspended and attached growth systems *Water Research* 47(2):623-36.

Tengerdy RP, Johnson JE, Holló J, Tóth J (1981) Denitrification and removal of heavy metals from waste water by immobilized microorganisms *Applied Biochemistry and Biotechnology* 6(1):3-13.

Thompson ED, Nakata HM (1971) Reduction of activity of reduced nicotinamide adenine dinucleotide oxidase by divalent cations in cell-free extracts of Bacillus cereus *Journal Bacteriology* 105(2):494-7.

Vainshtein M, Kuschk P, Mattusch J, Vatsourina A, Wiessner A (2003) Model experiments on the microbial removal of chromium from contaminated groundwater *Water Research* 37(6):1401-5.

Van Nooten T, Springael D, Bastiaens L (2008) Positive impact of microorganisms on the performance of laboratory-scale permeable reactive iron barriers *International Journal of Environmental Science and Technology* 42(5):1680-6.

Vargas-García M, del C López MJ, Suárez-Estrella F, Moreno J (2012) Compost as a source of microbial isolates for the bioremediation of heavy metals: in vitro selection *Science of the Total Environment* 431:62-7.

Vatsouria A, Vainshtein M, Kuschk P, Wiessner ADK, Kaestner M (2005) Anaerobic co-reduction of chromate and nitrate by bacterial cultures of Staphylococcus epidermidis L-02 *Journal of Industrial Microbiology and Biotechnology* 32(9):409-14 E.

Vaxevanidou K, Papassiopi N, Paspaliaris I (2008) Removal of heavy metals and arsenic from contaminated soils using bioremediation and chelant extraction techniques *Chemosphere* 70(8):1329-37.

Velimirovic M, Simons Q, Bastiaens L (2014) Guar gum coupled microscale ZVI for in situ treatment of CAHs: continuous-flow column study *Journal of Hazardous Materials* 265:20-9.

Velmurugan P, Shim J, You Y, Choi S, Kamala-Kannan S, Lee KJ, Kim HJ, Oh BT (2010) Removal of zinc by live dead and dried biomass of Fusarium spp isolated from the abandoned-metal mine in South Korea and its perspective of producing nanocrystals *Journal of Hazardous Materials* 182(1-3):317-24.

Vena MP, Jobbágy M, Bilmes SA (2016) Microorganism mediated biosynthesis of metal chalcogenides; a powerful tool to transform toxic effluents into functional nanomaterials *Science of the Total Environment* 565:804-810.

Villegas LB, Fernández PM, Amoroso MJ, de Figueroa LI (2008) Chromate removal by yeasts isolated from sediments of a tanning factory and a mine site in Argentina *Biometals* 21(5):591-600.

Vymazal J (2005) Removal of heavy metals in a horizontal sub-surface flow constructed wetland *Journal of Environmental Science and Health* Part A Toxic/hazardous substances and environmental engineering 40(6-7):1369-79.

Wadgaonkar SL, Nancharaiah YV, Esposito G Lens PNL (2018) Environmental impact and bioremediation of seleniferous soils and sediments *Critical Reviews in Biotechnology* 38(6):941-956.

Wagner FB, Nielsen P, Boe-Hansen R, Albrechtsen HJ (2016) Copper deficiency can limit nitrification in biological rapid sand filters for drinking water production *Water Research* 95:280-8.

Wang L, Yang C, Cheng Y, Huang J, He H, Zeng G, Lu L (2013) Effects of surfactant and Zn (II) at various concentrations on microbial activity and ethylbenzene removal in biotricking filter *Chemosphere* 93(11):2909-13.

Wang MY, Ma LM (2014) [Investigation of enhanced low carbon wastewater denitrification by catalytic iron] *Huan Jing Ke Xue* 35(7):2633-8.

Wang N, Qiu Y, Xiao T, Wang J, Chen Y, Xu X, Kang Z, Fan L, Yu H (2019) Comparative studies on Pb(II) biosorption with three spongy microbe-based biosorbents: High performance selectivity and application *Journal of Hazardous Materials* 373:39-49.

Wang S, Li Z, Gao M, She Z, Ma B, Guo L, Zheng D, Zhao Y, Jin C, Wang X, Gao F (2017) Long-term effects of cupric oxide nanoparticles (CuO NPs) on the performance microbial community and enzymatic activity of activated sludge in a sequencing batch reactor *Journal of Environmental Management* 187:330-339.

Wang W, Zhang X, Huang J, Yan C, Zhang Q, Lu H, Liu J (2014) Interactive effects of cadmium and pyrene on contaminant removal from co-contaminated sediment planted with mangrove Kandelia obovata (S L) Yong seedlings *Marine Pollution Bulletin* 84(1-2): 306-13.

Wang X He Z Luo H Zhang M Zhang D Pan X Gadd G M (2018) Multiple-pathway remediation of mercury contamination by a versatile selenite-reducing bacterium *Science of the Total Environment* 615: 615-623.

Wang Y, Li J, Zhai S, Wei Z, Feng J (2015) Enhanced phosphorus removal by microbial-collaborating sponge iron *Water Science and Technology* 72(8):1257-65.

Wang Y, Qin J, Zhou S, Lin X, Ye L, Song C, Yan Y (2015) Identification of the function of extracellular polymeric substances (EPS) in denitrifying phosphorus removal sludge in the presence of copper ion *Water Research* 73:252-64.

Wang Z, Gao M, She Z, Jin C, Zhao Y, Yang S, Guo L, Wang S (2015) Effects of hexavalent chromium on performance and microbial community of an aerobic granular sequencing batch reactor *Environmental Science and Pollution* 22(6):4575-86.

Watanabe T, Jin HW, Cho KJ, Kuroda M (2004) Application of a bio-electrochemical reactor process to direct treatment of metal pickling wastewater containing heavy metals and high strength nitrate *Water Science and Technology* 50(8):111-8.

Watanabe T, Motoyama H, Kuroda M (2001) Denitrification and neutralization treatment by direct feeding of an acidic wastewater containing copper ion and high-strength nitrate to a bio-electrochemical reactor process *Water Research* 35(17):4102-10.

Wei J, Liu X, Wang Q, Wang C, Chen X, Li H (2014) Effect of rhizodeposition on pyrene bioaccessibility and microbial structure in pyrene and pyrene-lead polluted soil *Chemosphere* 97:92-7.

Williams KH, Wilkins MJ, N'Guessan AL, Arey B, Dodova E, Dohnalkova A, Holmes D Lovley DR, Long PE (2013) Field evidence of selenium bioreduction in a uranium-contaminated aquifer *Environmental Microbiology Reports* 5(3):444-52.

Windler L, Height M, Nowack B (2013) Comparative evaluation of antimicrobials for textile applications *Environment International* 53:62-73.

Wu D, Shen Y, Ding A, Mahmood Q, Liu S, Tu Q (2013) Effects of nanoscale zero-valent iron particles on biological nitrogen and phosphorus removal and microorganisms in activated sludge *Journal of Hazardous Materials* 262:649-55.

Wu X, Li W, Ou D, Li C, Hou M, Li H, Liu Y (2019) Enhanced adsorption of Zn2+ by salinity-aided aerobic granular sludge: Performance and binding mechanism *Journal of Environmental Management* 242:266-271.

Wu X, Tong F,, Yong X Zhou J,, Zhang L Jia H, Wei P (2016) Effect of NaX zeolite-modified graphite felts on hexavalent chromium removal in biocathode microbial fuel cells *Journal of Hazardous Materials* 308:303-11.

Wu Y, Wang L, Jin M, Kong F, Qi H, Nan J (2019) Reduced graphene oxide and biofilms as cathode catalysts to enhance energy and metal recovery in microbial fuel cell *Bioresource Technology* 283:129-137.

Xie Y, Li X, Huang X, Han S, Amombo E, Wassie M, Chen L, Fu J (2019) Characterization of the Cd-resistant fungus Aspergillus aculeatus and its potential for increasing the antioxidant activity and photosynthetic efficiency of rice *Ecotoxicology and Environmental Safety* 171:373-381.

Xu YB, Xiao HH, Sun SY (2005) Study on anaerobic treatment of wastewater containing hexavalent chromium *Journal of Zhejiang University Science* B 6(6):574-9.

Xu X, Xia S, Zhou L, Zhang Z, Rittmann B E (2015) Bioreduction of vanadium (V) in groundwater by autohydrogentrophic bacteria: Mechanisms and microorganisms *Journal of Environmental Sciences* (China) 30:122-8.

Yan R, Yang F, Wu Y, Hu Z, Nath B, Yang L, Fang Y (2011) Cadmium and mercury removal from non-point source wastewater by a hybrid bioreactor *Bioresource Technology* 102(21):9927-32.

Yang S, Shibata A, Yoshida N, Katayama A (2009) Anaerobic mineralization of pentachlorophenol (PCP) by combining PCP-dechlorinating and phenol-degrading cultures *Biotechnology and Bioengineering* 102(1):81-90.

Yang Y, Inamori Y, Ojima H, Machii H, Shimizu Y (2005) Development of an advanced biological treatment system applied to the removal of nitrogen and phosphorus using the sludge ceramics *Water Research* 39(20):4859-68.

Yang L Li X Chu Z Ren Y Zhang J (2014) Distribution and genetic diversity of the microorganisms in the biofilter for the simultaneous removal of arsenic iron and manganese from simulated groundwater *Bioresource Technology* 156:384-8.

Ye M, Sun M, Liu Z, Ni N, Chen Y, Gu C, Kengara FO, Li H, Jiang X (2014) Evaluation of enhanced soil washing process and phytoremediation with maize oil carboxymethyl-β-cyclodextrin and vetiver grass for the recovery of organochlorine pesticides and heavy metals from a pesticide factory site *Journal of Environmental Management* 141:161-8.

Ye M, Sun M, Wan J, Fang G, Li H, Hu F, Jiang X, Kengara F O (2015) Enhanced soil washing process for the remediation of PBDEs/Pb/Cd-contaminated electronic waste site with carboxymethyl chitosan in a sunflower oil-water solvent system and microbial augmentation *Environmental Science and Pollution Research* 22(4):2687-98.

Yin K, Yang Z, Gong Y, Wang D, Lin H (2019) The antagonistic effect of Se on the Pb-weakening formation of neutrophil extracellular traps in chicken neutrophils *Ecotoxicology and Environmental Safety* 173:225-234.

Yin W, Li Y, Wu J, Chen G, Jiang G, Li P, Gu J, Liang H, Liu C (2017) Enhanced Cr(VI) removal from groundwater by Fe0-H2O system with bio-amended iron corrosion *Journal of Hazardous Materials* 332:42-50.

Yin W, Wu J, Huang W, Li Y, Jiang G (2016) The effects of flow rate and concentration on nitrobenzene removal in abiotic and biotic zero-valent iron columns *Science of the Total Environment* 560-561:12-8.

Yoon KY, Byeon JH, Park CW, Hwang J (2008) Antimicrobial effect of silver particles on bacterial contamination of activated carbon fibers *Environmental Science and Technology* 42(4):1251-5.

Yu X, Amrhein C, Deshusses MA, Matsumoto MR (2007) Perchlorate reduction by autotrophic bacteria attached to zerovalent iron in a flow-through reactor *Environmental Science and Technology* 41(3):990-7.

Yuan L, Zhi W, Liu Y, Karyala S, Vikesland PJ, Chen X, Zhang H (2015) Lead toxicity to the performance viability and community composition of activated sludge microorganisms *International Journal of Environmental Science and Technology* 49(2):824-30.

Zelepukha SI, Zemlerub IL, Zaiakin SR (1975) Experimental removal of microorganisms from water by the contact coagulation method] Mikrobiol Zh 37(5):648-52.

Zhang B, Li F, Houk RS, Armstrong DW (2003) Pore exclusion chromatography-inductively coupled plasma-mass spectrometry for monitoring elements in bacteria: a study on microbial removal of uranium from aqueous solution *Analytical Chemistry* 75(24):6901-5.

Zhang Z, Rengel Z, Meney K, Pantelic L, Tomanovic R (2011) Polynuclear aromatic hydrocarbons (PAHs) mediate cadmium toxicity to an emergent wetland species *Journal of Hazardous Materials* 189(1-2):119-26.

Zhang B, Hao L, Tian C, Yuan S, Feng C, Ni Jk, Borthwick AG (2015) Microbial reduction and precipitation of vanadium (V) in groundwater by immobilized mixed anaerobic culture *Bioresource Technology* 192:410-7.

Zhang D, Trzcinski AP, Oh HS, Chew E, Liu Y, Tan SK, Ng WJ (2017) Comparison of the effects and distribution of zinc oxide nanoparticles and zinc ions in activated sludge reactors *Journal of Environmental Science and Health A Tox Hazard Subst Environ Eng* 52(11):1073-1081.

Zhang D, Trzcinski AP, Oh HS, Chew E, Tan SK, Ng W, J Liu Y (2017) Comparison and distribution of copper oxide nanoparticles and copper ions in activated sludge reactors *J Environ Sci Health A Tox Hazard Subst Environ Eng* 52(6):507-514.

Zhang K, Xue Y, Xu H, Yao Y (2019) Lead removal by phosphate solubilizing bacteria isolated from soil through biomineralization *Chemosphere* 224:272-279.

Zhang L, Lin X, Wang J, Jiang F, Wei L, Chen G, Hao X (2016) Effects of Lead and Mercury on Sulfate-Reducing Bacterial Activity in a

Biological Process for Flue Gas Desulfurization Wastewater Treatment *Scientific Reports* 6:30455.

Zhang R, Huang T, Wen G, Chen Y, Cao X, Zhang B (2017) Using Iron-Manganese Co-Oxide Filter Film to Remove Ammonium from Surface Water *International Journal of Environmental Research and Public Health* 14(7)

Zhang Y, Douglas GB, Kaksonen AH, Cui L, Ye Z (2019) Microbial reduction of nitrate in the presence of zero-valent iron *Science of the Total Environment* 646:1195-1203.

Zhang Y, Li X, Fu C, Li X, Yan B, Shi T (2019) Effect of Fe^{2+} addition on chemical oxygen demand and nitrogen removal in horizontal subsurface flow constructed wetlands *Chemosphere* 220:259-265.

Zhang Y, Li G, Wen J, Xu Y, Sun J, Ning XA, Lu X, Wang Y, Yang Z, Yuan Y (2018) Electrochemical and microbial community responses of electrochemically active biofilms to copper ions in bioelectrochemical systems *Chemosphere* 196:377-385.

Zhang Y, Liu X, Fu C, Li X, Yan B, Shi T (2019) Effect of Fe^{2+} addition on chemical oxygen demand and nitrogen removal in horizontal subsurface flow constructed wetlands *Chemosphere* 220:259-265.

Zhen G, Lu X, Kobayashi T, Su L, Kumar G, Bakonyi P, He Y, Sivagurunathan P, Nemestóthy N, Xu K, Zhao Y (2017) Continuous micro-current stimulation to upgrade methanolic wastewater biodegradation and biomethane recovery in an upflow anaerobic sludge blanket (UASB) reactor *Chemosphere* 180:229-238.

Zheng K, Fan J, Hu X, Zhang X, Liu X, Shen J (2019) Distribution by influence factors of pyrene removal in chemical enhancers assisted microbial phytoremediation of Scirpus triqueter in co-contaminated soils *Journal International Journal of Phytoremediation* 1-7.

Zheng XY, Zhu X, Xu YD, Wang J, Wei C, Gao YJ, Zhou G (2017) [Removal of Nitrogen in Municipal Secondary Effluent by a Vertical Flow Constructed Wetland Associated with Iron-carbon *Internal Electrolysis*] *Huan Jing Ke Xue* 38(6):2412-2418.

Zhong J, Yin W, Li Y, Li P, Wu J, Jiang G, Gu J, Liang H (2017) Column study of enhanced Cr(VI) removal and longevity by coupled abiotic and biotic processes using Fe0 and mixed anaerobic culture *Water Research* 122:536-544.

Zhou LC, Li YF, Bai X, Zhao GH (2009) Use of microorganisms immobilized on composite polyurethane foam to remove Cu(II) from aqueous solution *Journal of Hazardous Materials* 167(1-3):1106-13.

Zuo Y, Chen G, Zeng G, Li Z, Yan M, Chen A, Guo Z, Huang Z, Tan Q (2015) Transport fate and stimulating impact of silver nanoparticles on the removal of Cd(II) by Phanerochaete chrysosporium in aqueous solutions *Journal of Hazardous Materials* 285:236-44.

In: Environmental Science of Heavy Metals ISBN: 978-1-53617-831-9
Editor: Dorota Bartusik-Aebisher © 2020 Nova Science Publishers, Inc.

Chapter 2

PHOTOACTIVE MATERIALS FOR HEAVY METAL REMOVAL

*David Aebisher, Sabina Galiniak, Tomasz Kubrak,
Rafał Podgórski and Dorota Bartusik-Aebisher*[*]
Faculty of Medicine, University of Rzeszów

The pollution of water by heavy metals is a visibly growing problem. The importance of natural water in rivers, streams, seas, and oceans is enormous, and cannot be understated. Water is the primary component of the human body, animals and plants and is the environment of many species of animals, microorganisms and plants. Water is also used in industry and is an indispensable material for human hygiene. Waterways are used for communication and transport. The development of new water purification agents based on adsorbents is required. One such material is based on silica with chemical modification with porphyrins or porphyrin derivatives. Both derivatives of porphyrin and porhyrin attached to silica are designed to be new photoactive materials and are being recognized for

---

[*] Corresponding Author's Email: Dbartusik-aebisher@ur.edu.pl.

heavy metal removal (Pelza et al., 2017). A primary target of purification is wastewater, which is a mixture of household waste, feces, and waste from hospitals, baths, laundries and industrial plants. A significant part of wastewater is suspended or dissolved organic compounds, mainly proteins, fats and carbohydrates. It also contains detergents, and pathogenic microorganisms that are the source of such diseases as typhus, cholera, typhoid fever, and Hein's Medina disease. Heavy metals (lead, mercury) are another important and toxic component of wastewater. These substances, when ingested by animals or humans, cause damage to the liver, blood vessels, heart, nervous system, and bones.

Recently, photocatalytic mineral paints have been used outdoors to reduce atmospheric pollution such as nitrogen oxides, greenhouse gases or surface pollution. Photoactive materials based on porphyrins possess novel properties based on their size, shape, chemical composition, and crystal structure (Kim et al., 2018; Smith et al., 2013). Porphyrin derivatives such as carboxy-, phenoxy-, pyridyl-, and dimethoxy- compounds are well known and are obtained by synthesis some of which have been characterized as ionophores for preparing membrane sensors selective to iron (III) (Vlascici et al., 2012; Wu et al., 2017). Photosensitizers are substances that absorb radiant energy, especially visible light, and transfer energy the environment. They are used to give photosensitive properties to plastics (e.g., for the production of printing plates), as well as to destroy them after some time of exposure. Promising photoactive conjugated polymers have attracted extensive attention as metal-free and visible-light-responsive photocatalysts in the area of purification of water (Zheng et al., 2017). Calcined Zn-Al layered double hydroxides and organo-modified double hydroxides were applied as adsorbents. Both materials are photocatalysts, and their main application was for removal of cationic dyes (Starukh 2017). Tuning and optimizing the efficiency of light energy transfer plays an important role in meeting modern challenges of minimizing energy loss and developing high-performance optoelectronic materials (Gao et al., 2017; Han et al., 2017). In medicine, photodynamic therapy (PDT) uses visible light and dye to produce reactive oxygen species that can kill cancer cells and infectious microorganisms (Huang et al., 2012). The fabrication of industrial-scale photoreactive composite materials containing living cells has been achieved (Jenkins et al., 2013; Venditti 2017; Lim et al., 2016; Sharma et al., 2012; Gao et al., 2015; Reynoso et al., 2019; Kadhom et al., 2017; Cabello et al., 2018). Material, acting as photoactive sensors, are a major component of chemical biology (Haas et al., 2009). Nanomaterials based on porphyrins have become the focus of recent material development (Petersen et al., 2014). Porous materials such as zeolites containing porphyrin derivatives are new established classes of compounds (Tansell et al., 2017). For example, azobenzene derivatives, due to their photo- and electroactive properties, are an important group of compounds finding applications in diverse fields (Wagner-Wysiecka et al., 2018; Rezaei et al., 2018). Fruitful efforts have been made in combining molecular technology with biosensing (Runfa et al., 2019; Barroso-Martín et al., 2018; Shaik et al., 2019; Guerra et al., 2018; Drewniak et al., 2015). Materials for photoactive sensors are a major component of recent research in photobiology and photochemistry (Knoblauch et al., 2014). Materials for sensors used in science have seen a great deal of advancement and development (Chedid et al., 2018; Ashraf et al., 2019; Dalle et al., 2019). The surface characteristics of these sensors have been well analyzed (Liang et al., 2018; Faizan et al., 2019; Mesquita et al., 2018; Grayson et al., 2018; Zhou et al., 2018; Sobhanadhas et al., 2018; Babayigit et al., 2016; Perkins et al., 2006; Lünsdorf et al., 1997). Nanoparticles have recently attracted attention as a new class of compounds for cleaning the enviroment (Zemskova et al., 2018; Xiang et al., 2017; Liu et al., 2017; Cavallo et al., 2016; Hayyan et al., 2015; Litzov et al., 2013). New materials have applications in many fields such as biology, medicine, catalysis, nanoscience, redox, and photoactive materials (Wei et al., 2018; Baeza et al., 2018; Ligon et al., 2017; Min et al., 2016; Dragutan et al., 2015; Nayak et al., 2015; Heinemann et al., 2014). In many aspects, nanoporous carbons are similar to graphene; their pores are built of distorted graphene layers and defects arise from their amorphicity enabling reactivity (Hlapisi et al., 2019; Bandosz et al., 2018; Zha et al., 2018). Membrane development has been a major issue in the development of more efficient water treatment processes (Coelho et al., 2019; Dannert et al., 2019; Feng et al., 2019; Wen et al., 2017; Klang et al., 2013; Mändl 2009). Many wastes are a potential source of environment contamination due to inproper disposal (Tummino et al., 2019; Reheman et al., 2018; Petrova et al., 2012; Brown et al., 2018; Sutar et al., 2018; Maksimov et al., 2017; Bekkali et al., 2016; Coe et al., 2016; Qiu et al., 2013). The engineering of new materials with optical and acoustic functionalities is an indispensable and powerful means to develop novel materials for heavy metal removal (Lee et al., 2014; Gupta et al., 2012; Hedley et al., 2017; Kasyanenko et al., 2017; Li et al., 2017; Liu et al., 2016;Singh et al., 2015;Vernooij et al., 2018; Karcz et al., 2014). Photoconversion is a technique by which a fluorescent dye undergoes a shift in fluorescence following light absorption (Malatesta et al., 2013). Synergic behavior is always inspiring scientists to develop advanced material nanocomposites with superior performance and novel properties (Sheheri et al., 2019). Photocatalytically active nanostructures require a large specific surface area with the presence of many active sites for oxidation and reduction half reactions, and fast electron diffusion and charge separation (Maijenburg et al., 2014). Table 1 presents various technologies used for heavy metal removal. By combining photocatalytically active dyes with mineral binders, paint that combines color fastness and high dirt resistance was designed. Compared to conventional organic paints (such as emulsion paints or based on silicone resins), photocatalytic mineral paints offer real added value as cleaning materials.

Nanotechnology with photosensitizers is an evolving field with enormous potential for biomedical applications (Bhattacharyya et al., 2011; Cheng et al., 2015; Donaldson et al., 2016; Kugler et al., 2019; Dawod et al., 2018; Poplata et al., 2016; Mondal et al., 2015; Tan et al., 2013). The molecular interfacing of biomolecules with advanced materials, such as semiconductor nanocrystals, carbon nanotubes, and conducting polymers, has enormous potential in the field of biomedicine and optoelectronics (Iyer et al., 2007; Schweimer et al., 2000; Mishra et al., 2019; Dancer 2014; Takashima et al., 2011; Dancer 2014). Light is collected by pigment-protein complexes, which funnel solar energy at the start of photosynthesis (De Vico et al., 2018; Service et al., 2011; Golovynskyi et al., 2017). The ability of living systems to respond to stimuli and process information has encouraged scientists to develop integrated nanosystems displaying similar functions and capabilities (Pérez-Mitta et al., 2017; Hollingsworth et al., 2016; Batyuk et al., 2016; Kutta et al., 2015; Yang et al., 2015, Tallini et al., 2006). Nanostructured semiconductors feature resonant optical modes that confine light absorption in specific areas (Kontoleta et al., 2018).

Table 1. Various technologies used for heavy metal removal

Authors	Technology
Su et al., 2018; Janek et al., 2018; McGuire et al., 2018; Vega-Figueroa et al., 2018; Jones et al., 2017; Taabache et al., 2017; Nedosekin et al., 2017; Karimi et al., 2017; Nishima et al., 2016; Tuo et al., 2015; Walter et al., 2015; Sun et al., 2015; Hwang et al., 2014; Jackson et al., 2013; Bian et al., 2012; Liu et al., 2009; Makarava et al., 2005; Edelman 1990; Butler et al., 1984; Butler et al., 1980	nanotechnology
Green wt al., 2019; Chu et al., 2019; Kimmel et al., 2011; Jorge et al., 2007; Azizi et al., 2017; Chakraborty et al., 2017; Toth et al., 2015;; Satori et al., 2013; Prakadan et al., 2017; Luo et al., 2015; Jjemba et al., 2015	metal coridanation chemistry
Zou et al., 2018; Cho et al., 2018; Xia et al., 2017; Han et al., 2017; Denisov et al., 2017; Rajasekaran et al., 2017; Laysandra et al., 2017; Paz 2011; Razzaq et al., 2019; Zhang et al., 2015; Carré et al., 2014; (Riedel et al., 2013 Xu et al., 2013; Rasmussen et al., 2018; Liu et al., 2018; Suskind et al., 1977	photochemical transformations relevant to saturated carbonyls

Intrabodies enable targeting of proteins in live cells, but generating specific intrabodies against the thousands of proteins in a proteome poses a challenge (Prole et al., 2019; Evison et al., 2016, Fullagar et al., 2017; Hansel et al., 2006; Chakraborty et al., 2017; Wang et al., 2011). Fluorophores that absorb and emit in the red spectral region (600-700 nm) are of great interest in photochemistry and photomedicine (Krivanek et al., 2008; Liu et al., 2018; Bazzan et al., 2008). Gold nanoparticles used **as sensors** have been widely employed in various fields including biology, environmental sensors and food toxin detection, but their sensitivity to size- and shape-dependent activity limits their practical applications (Song et al., 2017). Nanoparticles are used in techniques suitable to quantify material uptake by tissue and cells. The most commonly applied techniques for this purpose are based on inductively coupled plasma mass spectrometry (Hsiao et al., 2016; Hoffman-Luca et al., 2009). The kinetics of photoinduced absorbance changes in the 400 ns to 100 ms time range were studied between 770 and 1025 nm in reaction center core complexes isolated from the green sulfur bacterium Chlorobium vibrioforme (Vassiliev et al., 2001). Novel heteroleptic iridium complexes containing the 1-substituted-4-phenyl-1H-1,2,3-triazole cyclometalating ligand have been synthesized (Felici et al., 2010). Baker's yeast, S. cerevisiae, is a model organism that is used in synthetic biology. Work has demonstrated how nanostructured thin films can encode physiological responses in S. cerevisiae yeast (Snyder et al., 2018). Nanoporous alumina membranes exhibit high pore densities, well-controlled and uniform pore sizes, as well as straight pores (Narayan et al., 2010). Conventional cancer detection techniques show several limitations including low to no specificity and consequently low efficacy in discriminating between cancer cells and healthy cells (Jahangirian et al., 2019). Indium use has increased greatly in the past decade in parallel with the growth of flat-panel displays, touchscreens, optoelectronic devices, and photovoltaic cells. Much of this growth has been in the use of indium tin oxide (Hines et al., 2013). A simple synthesis route for growth of Ag/AgO nanoparticles in large quantitative yields with narrow size distribution from a functional, non-activated, Ni (II) based highly flexible porous coordination polymer as a template has been demonstrated (Agarwal et al., 2017). Oligonucleotides carrying thiol groups are useful intermediates for a remarkable number of applications involving nucleic acids. In one study, DNA oligonucleotides carrying tert-butylsulfanyl protected thiol groups have been prepared (Pérez-Rentero et al., 2012). Histidine tags which are commonly used for purification of recombinant proteins were converted to a catalytic redox-active center by incorporation of Co($^{2+}$). Two examples of the biological activity of this engineered protein-derived cofactor have been published (Shin et al., 2014; Lou et al., 2013). There is an increasing interest in the application of photocatalytic properties for disinfection of surfaces, air, and water (Le et al., 2011; Dickerson et al., 2008). To realize useful control over molecular motion in the future, an extensive toolbox of both actionable molecules and stimuli-responsive units must be developed (Findlay et al., 2018; Suyatin et al., 2013; Dickinson et al., 2013; Whited et al., 2009: Mackay et al., 2007; Kang et al., 2007). Uranium (VI) is considered to be one of the most widely dispersed and problematic environmental contaminants, due in large part to its high solubility and great mobility in natural aquatic systems (Rudack et al., 2015; Jiang et al., 2013; Burke et al., 2008; Stefanelli et al., 2014). Photodynamic properties and pharmacokinetics of dyes for many diverse applications have been developed (Eisenberger et al., 1982; Reddi et al., 1990; We Calvo et al., 1990; Thomson et al., 1990). For example, a covalently linked organic dye-cobaloxime catalyst system based on mesoporous NiO was synthesized by a facile click reaction for mechanistic studies and application in a dye-sensitized solar fuel device (Pati et al., 2017). Photochemical ribonucleotide reductases have been developed to study the proton-coupled electron transfer mechanism of radical transport in class I E. coli ribonucleotide reductase (Reece et al., 2009). Sensory rhodopsin II is a negative phototaxis receptor of Halobacterium salinarum, a bacterium that avoids blue-green light. In one study, the photoinduced proton transfer activity of rhodopsin II from Halobacterium salinarum was presented (Tamogami et al., 2010). Chemical pollutants are various dissolved inorganic or organic chemicals that get into natural waters together with wastewater from industrial plants. Many of them are toxic to living organisms: these include heavy metal salts, cyanides, phenols, and petroleum components. Some have an inhibitory effect on the natural biological processes occurring in water, or are undesirable due to deterioration of taste quality.

ACKNOWLEDGMENTS

Dorota Bartusik-Aebisher acknowledges support from the National Center of Science NCN (New drug delivery systems-MRI study, Grant OPUS-13 number 2017/25/B/ST4/02481).

REFERENCES

Alejandro Baeza & María Vallet-Regí. (2018). Nanomotors for Nucleic Acid, Proteins, Pollutants and Cells Detection. *Int J Mol Sci.*, *19*(6), 1579.

Alexander Batyuk, Lorenzo Galli, Andrii Ishchenko, Gye Won Han, Cornelius Gati, Petr A. Popov, Ming-Yue Lee, Benjamin Stauch, Thomas A. White, Anton Barty, Andrew Aquila, Mark S. Hunter, Mengning Liang, Sébastien Boutet, Mengchen Pu, Zhi-jie Liu, Garrett Nelson, Daniel James, Chufeng Li, Yun Zhao, John C. H. Spence, Wei Liu, Petra Fromme, Vsevolod Katritch, Uwe Weierstall, Raymond C. Stevens & Vadim Cherezov. (2016). Native phasing of x-ray free-electron laser data for a G protein–coupled receptor. *Sci Adv.*, *2*(9), e1600292.

Alexander J. Tansell, Corey L. Jones & Timothy L. Easun. (2017). MOF the beaten track: unusual structures and uncommon applications of metal–organic frameworks. *Chem Cent J.*, *11*, 100.

Allister F. McGuire, Francesca Santoro & Bianxiao Cui. (2018). Interfacing cells with vertical nanoscale devices: applications and characterization. *Annu Rev Anal Chem (Palo Alto Calif).*, *11*(1), 101–126.

Amy M. Wen & Nicole F. Steinmetz. (2017). Design of virus-based nanomaterials for medicine, biotechnology, and energy. *Chem Soc Rev.*

Ann-Karin Olsen, Nur Duale, Magnar Bjørås, Cathrine T. Larsen, Richard Wiger, Jørn A. Holme, Erling C. Seeberg & Gunnar Brunborg. (2003). Limited repair of 8-hydroxy-7,8-dihydroguanine residues in human testicular cells. *Nucleic Acids Res.*, *31*(4), 1351–1363.

Arati Sharma, SubbaRao V. Madhunapantula & Gavin P. Robertson. (2012). Toxicological considerations when creating nanoparticle based drugs and drug delivery systems? *Expert Opin Drug Metab Toxicol.*, *8*(1), 47–69.

Asheesh Gupta, Pinar Avci, Magesh Sadasivam, Rakkiyappan Chandran, Nivaldo Parizotto, Daniela Vecchio, Wanessa C Antunes-Melo,

Tianhong Dai, Long Y. Chiang & Michael R. Hamblin. (2012). Shining Light on Nanotechnology to Help Repair and Regeneration. *Biotechnol Adv.*

Aslihan Babayigit, Dinh Duy Thanh, Anitha Ethirajan, Jean Manca, Marc Muller, Hans-Gerd Boyen & Bert Conings. (2016). Assessing the toxicity of Pb- and Sn-based perovskite solar cells in model organism Danio rerio. *Sci Rep.*, *6*, 18721.

Athanasios A. Tountas, Xinyue Peng, Alexandra V. Tavasoli, Paul N. Duchesne, Thomas L. Dingle, Yuchan Dong, Lourdes Hurtado, Abhinav Mohan, Wei Sun, Ulrich Ulmer, Lu Wang, Thomas E. Wood, Christos T. Maravelias, Mohini M. Sain & Geoffrey A. Ozin. (2019). Towards Solar Methanol: Past, Present, and Future. *Adv Sci (Weinh).*, *6*(8), 1801903.

Award Winners and Abstracts of the 31st Annual Symposium of The Protein Society, Montreal, Canada, 24–27, *Protein Sci.*, *26*, (Suppl 1), 6–209.

Baiyan Li, Xinglong Dong, Hao Wang, Dingxuan Ma, Kui Tan, Stephanie Jensen, Benjamin J. Deibert, Joseph Butler, Jeremy Cure, Zhan Shi, Timo Thonhauser, Yves J. Chabal, Yu Han & Jing Li. (2017). Capture of organic iodides from nuclear waste by metal-organic framework-based molecular traps. *Nat Commun.*, *8*, 485.

Baltimore, M. D. (2016). Award Winners and Abstracts of the 30th Anniversary Symposium of The Protein Society. *Protein Sci.*, *25*, (Suppl 1), 4–176.

Beata Bajorowicz, Anna Cybula, Michał J. Winiarski, Tomasz Klimczuk & Adriana Zaleska. (2014). Surface Properties and Photocatalytic Activity of KTaO3, CdS, MoS2 Semiconductors and Their Binary and Ternary Semiconductor Composites. *Molecules.*, *19*(9), 15339–15360.

Beatriz Pelaz, Christoph Alexiou, Ramon A. Alvarez-Puebla, Frauke Alves, Anne M. Andrews, Sumaira Ashraf, Lajos P. Balogh, Laura Ballerini, Alessandra Bestetti, Cornelia Brendel, Susanna Bosi, Monica Carril, Warren C. W. Chan, Chunying Chen, Xiaodong Chen, Xiaoyuan Chen, Zhen Cheng, Daxiang Cui, Jianzhong Du, Christian Dullin, Alberto Escudero, Neus Feliu, Mingyuan Gao, Michael

George, Yury Gogotsi, Arnold Grünweller, Zhongwei Gu, Naomi J. Halas, Norbert Hampp, Roland K. Hartmann, Mark C. Hersam, Patrick Hunziker, Ji Jian, Xingyu Jiang, Philipp Jungebluth, Pranav Kadhiresan, Kazunori Kataoka, Ali Khademhosseini, Jindřich Kopeček, Nicholas A. Kotov, Harald F. Krug, Dong Soo Lee, Claus-Michael Lehr, Kam W. Leong, Xing-Jie Liang, Mei Ling Lim, Luis M. Liz-Marzán, Xiaowei Ma, Paolo Macchiarini, Huan Meng, Helmuth Möhwald, Paul Mulvaney, Andre E. Nel, Shuming Nie, Peter Nordlander, Teruo Okano, Jose Oliveira, Tai Hyun Park, Reginald M. Penner, Maurizio Prato, Victor Puntes, Vincent M. Rotello, Amila Samarakoon, Raymond E. Schaak, Youqing Shen, Sebastian Sjöqvist, Andre G. Skirtach, Mahmoud G. Soliman, Molly M. Stevens, Hsing-Wen Sung, Ben Zhong Tang, Rainer Tietze, Buddhisha N. Udugama, J. Scott VanEpps, Tanja Weil, Paul S. Weiss, Itamar Willner, Yuzhou Wu, Lily Yang, Zhao Yue, Qian Zhang, Qiang Zhang, Xian-En Zhang, Yuliang Zhao, Xin Zhou & Wolfgang J. Parak. (2017). Diverse Applications of Nanomedicine. *ACS Nano.*, *11*(3), 2313–2381.

Benjamin J. Evison, Marcelo L. Actis & Naoaki Fujii. (2016). A clickable psoralen to directly quantify DNA interstrand crosslinking and repair. *Bioorg Med Chem.*, *24*(5), 1071–1078.

Beverly A. Rzigalinski & Jeanine S. Strobl. (2009). Cadmium-Containing Nanoparticles: Perspectives on Pharmacology & Toxicology of Quantum Dots. *Toxicol Appl Pharmacol.*, *238*(3), 280–288.

Boyle, R. W., Rousseau, J., Kudrevich, S. V., Obochi, M. & van Lier, J. E. (1996). Hexadecafluorinated zinc phthalocyanine: photodynamic properties against the EMT-6 tumour in mice and pharmacokinetics using 65Zn as a radiotracer. *Br J Cancer.*, *73*(1), 49–53.

Butler, W. F., Calvo, R., Fredkin, D. R., Isaacson, R. A., Okamura, M. Y. & Feher, G. (1984). The electronic structure of Fe^{2+} in reaction centers from Rhodopseudomonas sphaeroides. III. EPR measurements of the reduced acceptor complex. *Biophys J.*, *45*(5), 947–973.

Butler, W. F., Johnston, D. C., Shore, H. B., Fredkin, D. R., Okamura, M. Y. & Feher, G. (1980). The electronic structure of Fe^{2+} in reaction centers from Rhodopseudomonas sphaeroides. I. Static magnetization measurements. *Biophys J.*, *32*(3), 967–992.

C. El Bekkali, H. Bouyarmane, S. Saoiabi, M. El Karbane, A. Rami, A. Saoiabi, M. Boujtita & Laghzizil, A. (2016). Low-cost composites based on porous titania–apatite surfaces for the removal of patent blue V from water: Effect of chemical structure of dye. *J Adv Res.*, *7*(6), 1009–1017.

Calvo, R., Passeggi, M. C., Isaacson, R. A., Okamura, M. Y. & Feher, G. (1990). Electron paramagnetic resonance investigation of photosynthetic reaction centers from Rhodobacter sphaeroides R-26 in which Fe^{2+} was replaced by Cu^{2+}. Determination of hyperfine interactions and exchange and dipole-dipole interactions between Cu^{2+} and QA-. *Biophys J.*, *58*(1), 149–165.

Chad P. Satori, Michelle M. Henderson, Elyse A. Krautkramer, Vratislav Kostal, Mark M. Distefano & Edgar A. Arriaga. (2013). Bioanalysis of eukaryotic organelles. *Chem Rev.*, *113*(4), 2733–2811.

Charlotte A. Whited, Wendy Belliston-Bittner, Alexander R. Dunn, Jay R. Winkler & Harry B. Gray. (2009). Nanosecond photoreduction of inducible nitric oxide synthase by a Ru-diimine electron tunneling wire bound distant from the active site. *J Inorg Biochem.*, *103*(6), 906–911.

Cheng-Li Wang, Weng-Sing Hwang, Kuo-Ming Chang, Horng-Huey Ko, Chi-Shiung Hsi, Hong-Hsin Huang & Moo-Chin Wang. (2011). Formation and Morphology of $Zn_2Ti_3O_8$ Powders Using Hydrothermal Process without Dispersant Agent or Mineralizer. *Int J Mol Sci.*, *12*(2), 935–945.

Chenkun Zhou, Haoran Lin, Yu Tian, Zhao Yuan, Ronald Clark, Banghao Chen, Lambertus J. van de Burgt, Jamie C. Wang, Yan Zhou, Kenneth Hanson, Quinton J. Meisner, Jennifer Neu, Tiglet Besara, Theo Siegrist, Eric Lambers, Peter Djurovich & Biwu Ma. (2018). Advances in Molecularly Imprinting Technology for Bioanalytical Applications. *Chem Sci.*, *9*(3), 586–593.

Christopher Knoblauch, Mark Griep & Craig Friedrich. (2014). Recent Advances in the Field of Bionanotechnology: An Insight into

Optoelectric Bacteriorhodopsin, Quantum Dots, and Noble Metal Nanoclusters. *Sensors (Basel).*, *14*(10), 19731–19766.

Colleen M. Hansel & Chris A. Francis. (2006). Coupled Photochemical and Enzymatic Mn(II) Oxidation Pathways of a Planktonic Roseobacter-Like Bacterium. *Appl Environ Microbiol.*, *72*(5), 3543–3549.

Corinna Dannert, Bjørn Torger Stokke & Rita S. Dias (2019). Nanoparticle-Hydrogel Composites: From Molecular Interactions to Macroscopic Behavior. *Polymers (Basel).*, *11*(2), 275.

Cynthia J. Hines, Jennifer L. Roberts, Ronnee N. Andrews, Matthew V. Jackson & James A. Deddens. (2013). *J Occup Environ Hyg. Use of and Occupational Exposure to Indium in the United States.*, *10*(12), 723–733.

Dana Vlascici, Eugenia Fagadar-Cosma, Iuliana Popa, Vlad Chiriac & Mayte Gil-Agusti. (2012). A Novel Sensor for Monitoring of Iron(III) Ions Based on Porphyrins. *Sensors (Basel).*, *12*(6), 8193–8203.

Daniel R. Jones, Virginia Gomez, Joseph C. Bear, Bertrand Rome, Francesco Mazzali, James D. McGettrick, Aled R. Lewis, Serena Margadonna, Waheed A. Al-Masry & Charles W. Dunnill. (2017). Active removal of waste dye pollutants using Ta3N5/W18O49 nanocomposite fibres. *Sci Rep.*, *7*, 4090.

Danielle W. Kimmel, Gabriel LeBlanc, Mika E. Meschievitz & David E. Cliffel. (2011). Electrochemical Sensors and Biosensors. Anal Chem.

Dariusz Karcz, Bożena Boroń, Arkadiusz Matwijczuk, Justyna Furso, Jakub Staroń, Alicja Ratuszna & Leszek Fiedor. (2014). Lessons from Chlorophylls: Modifications of Porphyrinoids Towards Optimized Solar Energy Conversion. *Molecules.*, *19*(10), 15938–15954.

David C. Green, Mark A. Holden, Mark A. Levenstein, Shuheng Zhang, Benjamin R. G. Johnson, Julia Gala de Pablo, Andrew Ward, Stanley W. Botchway & Fiona C. Meldrum. (2019). Controlling the fluorescence and room-temperature phosphorescence behaviour of carbon nanodots with inorganic crystalline nanocomposites. *Nat Commun.*, *10*, 206.

David F. Perkins, Leonard F. Lindoy, Alexander McAuley, George V. Meehan & Peter Turner. (2006). Manganese(II), iron(II), cobalt(II), and copper(II) complexes of an extended inherently chiral tris-bipyridyl cage. *Proc Natl Acad Sci U S A.*, *103*(3), 532–537.

David L. Prole & Colin W. Taylor. (2019). A genetically encoded toolkit of functionalized nanobodies against fluorescent proteins for visualizing and manipulating intracellular signaling. *BMC Biol.*, *17*, 41.

David W Manley, Andrew Mills, Christopher O'Rourke, Alexandra M Z Slawin & John C Walton. (2014). Catalyst-Free Photoredox Addition–Cyclisations: Exploitation of Natural Synergy between Aryl Acetic Acids and Maleimide. *Chemistry.*, *20*(18), 5492–5500.

Dmitry A. Nedosekin, Tariq Fahmi, Zeid A. Nima, Jacqueline Nolan, Chengzhong Cai, Mustafa Sarimollaoglu, Enkeleda Dervishi, Alexei Basnakian, Alexandru S. Biris & Vladimir P. Zharov. (2017). Photoacoustic flow cytometry for nanomaterial research. *Photoacoustics.*, *6*, 16–25.

Dmitry B. Suyatin, Lars Wallman, Jonas Thelin, Christelle N. Prinz, Henrik Jörntell, Lars Samuelson, Lars Montelius & Jens Schouenborg (2013). Nanowire-Based Electrode for Acute *In Vivo* Neural Recordings in the Brain. *PLoS One.*, *8*(2), e56673.

Dr. Palas Baran Pati, Dr. Lei Zhang, Dr. Bertrand Philippe, Ricardo Fernández-Terán, Dr. Sareh Ahmadi, Lei Tian, Prof. Dr. Håkan Rensmo, Prof. Dr. Leif Hammarström & Dr. Haining Tian. (2017). Insights into the Mechanism of a Covalently Linked Organic Dye–Cobaloxime Catalyst System for Dye-Sensitized Solar Fuel Devices. *ChemSusChem.*, *10*(11), 2480–2495.

Edelman, P. (1990). Environmental and workplace contamination in the semiconductor industry: implications for future health of the workforce and community. *Environ Health Perspect.*, *86*, 291–295.

Eduardo C. Reynoso, Eduardo Torres, Francesca Bettazzi & Ilaria Palchetti. (2019). Trends and Perspectives in Immunosensors for Determination of Currently-Used Pesticides: The Case of Glyphosate, Organophosphates, and Neonicotinoids. *Biosensors (Basel).*, *9*(1), 20.

Edward Tan, Feng Rao, Swathi Pasunooti, Thi Huong Pham, Ishin Soehano, Mark S. Turner, Chong Wai Liew, Julien Lescar, Konstantin Pervushin & Zhao-Xun Liang. (2013). Solution Structure of the PAS Domain of a Thermophilic YybT Protein Homolog Reveals a Potential Ligand-binding Site. *J Biol Chem.*, *288*(17), 11949–11959.

Eisenberger, P., Okamura, M. Y. & Feher, G. (1982). The electronic structure of Fe2+ in reaction centers from Rhodopseudomonas sphaeroides. II. Extended x-ray fine structure studies. *Biophys J.*, *37*(2), 523–538.

Elijah J. Petersen, Theodore B. Henry, Jian Zhao, Robert I. MacCuspie, Teresa L. Kirschling, Marina A. Dobrovolskaia, Vincent Hackley, Baoshan Xing & Jason C. White. (2014). Identification and Avoidance of Potential Artifacts and Misinterpretations in Nanomaterial Ecotoxicity Measurements. *Environ Sci Technol.*, *48*(8), 4226–4246.

Erin B. Dickerson, Erik C. Dreaden, Xiaohua Huang, Ivan H. El-Sayed, Hunghao Chu, Sujatha Pushpanketh, John F. McDonald & Mostafa A. El-Sayed. (2008). Gold nanorod assisted near-infrared plasmonic photothermal therapy (PPTT) of squamous cell carcinoma in mice. *Cancer Lett.*, *269*(1), 57–66.

Eugene G. Maksimov, Nikolai N. Sluchanko, Yury B. Slonimskiy, Kirill S. Mironov, Konstantin E. Klementiev, Marcus Moldenhauer, Thomas Friedrich, Dmitry A. Los, Vladimir Z. Paschenko & Andrew B. Rubin (2017). The Unique Protein-to-Protein Carotenoid Transfer Mechanism. *Biophys J.*, *113*(2), 402–414.

Eunah Kang, Jin-won Park, Scott McClellan, Jong-Mok Kim, David Holland, Gil U. Lee, Elias Franses, Kinam Park & David H. Thompson. (2007). Specific Adsorption of Histidine-Tagged Proteins on Silica Surfaces Modified with Ni2+:NTA-Derivatized Poly(Ethylene Glycol). *Langmuir.*, *23*(11), 6281–6288.

Evgenia Kontoleta, Sven H C Askes, Lai-Hung Lai & Erik C Garnett. (2018). Localized photodeposition of catalysts using nanophotonic resonances in silicon photocathodes. *Beilstein J Nanotechnol.*, *9*, 2097–2105.

Ewa Wagner-Wysiecka, Natalia Łukasik, Jan F. Biernat & Elżbieta Luboch. (2018). Azo group(s) in selected macrocyclic compounds. *J Incl Phenom Macrocycl Chem.*, *90*(3), 189–257. Correction in: *J Incl Phenom Macrocycl Chem.*, *90*(3), 259.

Fabrício Eduardo Bortot Coelho, Chiara Gionco, Maria Cristina Paganini, Paola Calza & Giuliana Magnacca. (2019). Control of Membrane Fouling in Organics Filtration Using Ce-Doped Zirconia and Visible Light. *Nanomaterials (Basel).*, *9*(4), 534.

Fatemeh Rezaei & Davide Vione. (2018). *Effect of pH on Zero Valent Iron Performance in Heterogeneous Fenton and Fenton-Like Processes: Molecules.*, *23*(12), 3127.

Fengjun Zha, Tingwei Wang, Ming Luo & Jianguo Guan. (2018). Tubular Micro/Nanomotors: Propulsion Mechanisms, Fabrication Techniques and Applications. *Micromachines (Basel).*, *9*(2), 78.

Fernanda D. Guerra, Mohamed F. Attia, Daniel C. Whitehead & Frank Alexis. (2018). Nanotechnology for Environmental Remediation: Materials and Applications. *Molecules.*, *23*(7), 1760.

Fiona S. Mackay, Julie A. Woods, Pavla Heringová, Jana Kašpárková, Ana M. Pizarro, Stephen A. Moggach, Simon Parsons, Viktor Brabec & Peter J. Sadler. (2007). A potent cytotoxic photoactivated platinum complex. *Proc Natl Acad Sci U S A.*, *104*(52), 20743–20748.

Gabriella Cavallo, Pierangelo Metrangolo, Roberto Milani, Tullio Pilati, Arri Priimagi, Giuseppe Resnati & Giancarlo Terraneo. (2016). The Halogen Bond. *Chem Rev.*, *116*(4), 2478–2601.

Gaëlle Carré, Erwann Hamon, Saïd Ennahar, Maxime Estner, Marie-Claire Lett, Peter Horvatovich, Jean-Pierre Gies, Valérie Keller, Nicolas Keller & Philippe Andre. (2014). TiO2 Photocatalysis Damages Lipids and Proteins in Escherichia coli. *Appl Environ Microbiol.*, *80*(8), 2573–2581.

Georges Chedid & Ali Yassin. (2018). Recent Trends in Covalent and Metal Organic Frameworks for Biomedical Applications. *Nanomaterials (Basel).*, *8*(11), 916.

Giorgio Bazzan, Wendy Smith, Lynn C. Francesconi & Charles Michael Drain. (2008). Electrostatic Self-Organization of Robust Porphyrin-Polyoxometalate Films. *Langmuir.*, *24*(7), 3244–3249.

Gong Cheng, Jing Lin, Jian Lu, Xi Zhao, Zhengqing Cai & Jie Fu. (2015). Advanced Treatment of Pesticide-Containing Wastewater Using Fenton Reagent Enhanced by Microwave Electrodeless Ultraviolet. *Biomed Res Int.*, *205903*.

Gonzalo Pérez-Mitta, Alberto G. Albesa, Christina Trautmann, María Eugenia Toimil-Molares & Omar Azzaroni. (2017). Bioinspired integrated nanosystems based on solid-state nanopores: "iontronic" transduction of biological, chemical and physical stimuli. *Chem Sci.*, *8*(2), 890–913.

Gopal Iyer, Fabien Pinaud, James Tsay & Shimon Weiss. (2007). *Solubilization of Quantum Dots with a Recombinant Peptide from Escherichia coli.*, *3*(5), 793–798.

Gordon J. Hedley, Arvydas Ruseckas & Ifor D. W. Samuel. (2017). Light Harvesting for Organic Photovoltaics. *Chem Rev.*, *117*(2), 796–837.

Gourav Mishra & Mausumi Mukhopadhyay. (2019). TiO2 decorated functionalized halloysite nanotubes (TiO2@HNTs) and photocatalytic PVC membranes synthesis, characterization and its application in water treatment. *Sci Rep.*, *9*, 4345.

Hai-Bing Xu, Peng-Chong Jiao, Bin Kang, Jian-Guo Deng & Yan Zhang. (2013). Walkable Dual Emissions. *Sci Rep.*, *3*, 2199.

Hong Su, Yafei Wang, Yuanliang Gu, Linda Bowman, Jinshun Zhao & Min Ding. (2018). Potential applications and human biosafety of nanomaterials used in nanomedicine. *J Appl Toxicol.*, *38*(1), 3–24.

Hossein Jahangirian, Katayoon Kalantari, Zahra Izadiyan, Roshanak Rafiee-Moghaddam, Kamyar Shameli & Thomas J Webster. (2019). A review of small molecules and drug delivery applications using gold and iron nanoparticles. *Int J Nanomedicine.*, *14*, 1633–1657.

Hui Liu, Shou-Wei Gao, Jing-Sheng Cai, Cheng-Lin He, Jia-Jun Mao, Tian-Xue Zhu, Zhong Chen, Jian-Ying Huang, Kai Meng, Ke-Qin Zhang, Salem S. Al-Deyab Yue-Kun Lai. (2016). Recent Progress in

&Fabrication and Applications of Superhydrophobic Coating on Cellulose-Based Substrates. *Materials (Basel).*, *9*(3), 124.

Hwang, M. S., Schwall, C. T., Pazarentzos, E., Datler, C., Alder, N. N. & Grimm, S. (2014). Mitochondrial Ca2+ influx targets cardiolipin to disintegrate respiratory chain complex II for cell death induction. *Cell Death Differ.*, *21*(11), 1733–1745.

Ibrahim, I., Lim, H. N., Huang, N. M. & Pandikumar, A. (2016). Cadmium Sulphide-Reduced Graphene Oxide-Modified Photoelectrode-Based Photoelectrochemical Sensing Platform for Copper(II) Ions. *PLoS One.*, *11*(5), e0154557.

Ileana Dragutan, Valerian Dragutan & Albert Demonceau. (2015). Editorial of Special Issue Ruthenium Complex: The Expanding Chemistry of the Ruthenium Complexes. *Molecules.*, *20*(9), 17244–17274.

Ilia G. Denisov & Stephen G. Sligar. (2017). Nanodiscs in Membrane Biochemistry and Biophysics. *Chem Rev.*, *117*(6), 4669–4713.

I-Lun Hsiao, Frank S. Bierkandt, Philipp Reichardt, Andreas Luch, Yuh-Jeen Huang, Norbert Jakubowski, Jutta Tentschert & Andrea Haase (2016). Quantification and visualization of cellular uptake of TiO2 and Ag nanoparticles: comparison of different ICP-MS techniques. *J Nanobiotechnology.*, *14*, 50.

In This Issue. (2007). *Proc Natl Acad Sci U S A.*, 104(51), 20145–20146.

Indranil Chakraborty, Jorge Jimenez & Mascharak, P. K. (2017). CO-Induced apoptotic death of colorectal cancer cells by a luminescent photoCORM grafted on biocompatible carboxymethyl chitosan. *Chem Commun (Camb).*, *53*(40), 5519–5522.

Indranil Chakraborty, Jorge Jimenez, W. M. C. Sameera, Masako Kato & Pradip K. Mascharak (2017). Luminescent Re(I) Carbonyl Complexes as Trackable PhotoCORMs for CO delivery to Cellular Targets. *Inorg Chem.*, *56*(5), 2863–2873.

Iole Venditti. (2017). Gold Nanoparticles in Photonic Crystals Applications: A Review. *Materials (Basel).*, *10*(2), 97.

Isabel Barroso-Martín, Elisa Moretti, Aldo Talon, Loretta Storaro, Enrique Rodríguez-Castellón & Antonia Infantes-Molina. (2018). Au and

AuCu Nanoparticles Supported on SBA-15 Ordered Mesoporous Titania-Silica as Catalysts for Methylene Blue Photodegradation. *Materials (Basel).*, *11*(6), 890.

Ivan Litzov & Christoph J. Brabec. (2013). Development of Efficient and Stable Inverted Bulk Heterojunction (BHJ) Solar Cells Using Different Metal Oxide Interfaces. *Materials (Basel).*, *6*(12), 5796–5820.

Jae-Hwang Lee, Cheong Yang Koh, Jonathan P Singer, Seog-Jin Jeon, Martin Maldovan, Ori Stein & Edwin L Thomas. (2014). 25th Anniversary Article: Ordered Polymer Structures for the Engineering of Photons and Phonons. *Adv Mater.*, *26*(4), 532–569.

James A. Findlay, Jonathan E. Barnsley, Keith C. Gordon & James D. Crowley. (2018). Synthesis and Light-Induced Actuation of Photo-Labile 2-Pyridyl-1,2,3-Triazole Ru(bis-bipyridyl) Appended Ferrocene Rotors. *Molecules.*, *23*(8), 2037.

James W. Dickinson, Michael Bromley, Fabrice P. L. Andrieux & Colin Boxall. (2013). Fabrication and Characterisation of the Graphene Ring Micro Electrode (GRiME) with an Integrated, Concentric Ag/AgCl Reference Electrode. *Sensors (Basel).*, *13*(3), 3635–3651.

Javoris V. Hollingsworth, Dinesh K. Bhupathiraju, N. V. S., Jirun Sun, Eric Lochner, M. Graça H. Vicente & Paul S. Russo. (2016). Preparation of Metalloporphyrin-Bound Superparamagnetic Silica Particles via "Click" Reaction. *ACS Appl Mater Interfaces.*, *8*(1), 792–801.

Jesse Coe & Petra Fromme. (2016). Serial Femtosecond Crystallography Opens New Avenues for Structural Biology. *Protein Pept Lett.*, *23*(3), 255–272.

Jessica S. Jenkins, Michael C. Flickinger & Orlin D. Velev. (2013). Engineering Cellular Photocomposite Materials Using Convective Assembly. *Materials (Basel).*, *6*(5), 1803–1825.

Jing Liu, Xiao-Min Li, Jing He, Lu-Ying Wang & Jian-Du Lei. (2018). Combining the Photocatalysis and Absorption Properties of Core-Shell Cu-BTC@TiO2 Microspheres: Highly Efficient Desulfurization of Thiophenic Compounds from Fuel. *Materials (Basel).*, *11*(11), 2209.

John E. Burke, Mark J. Karbarz, Raymond A. Deems, Sheng Li, Virgil L. Woods, Jr. & Edward A. Dennis. (2008). Interaction of Group IA Phospholipase A2 with Metal Ions and Phospholipid Vesicles Probed with Deuterium Exchange Mass Spectrometry. *Biochemistry.*, *47*(24), 6451–6459.

Jun Tamogami, Takashi Kikukawa, Yoichi Ikeda, Ayaka Takemura, Makoto Demura & Naoki Kamo. (2010). The Photochemical Reaction Cycle and Photoinduced Proton Transfer of Sensory Rhodopsin II (Phoborhodopsin) from Halobacterium salinarum. *Biophys J.*, *98*(7), 1353–1363.

Karlene Vega-Figueroa, Jaime Santillán, Valerie Ortiz-Gómez, Edwin O. Ortiz-Quiles, Beatriz A. Quiñones-Colón, David A. Castilla-Casadiego, Jorge Almodóvar, Marvin J. Bayro, José A. Rodríguez-Martínez & Eduardo Nicolau. (2018). Aptamer-Based Impedimetric Assay of Arsenite in Water: Interfacial Properties and Performance. *ACS Omega.*, *3*(2), 1437–1444.

Kathryn L. Haas & Katherine J. Franz. (2009). Application of Metal Coordination Chemistry to Explore and Manipulate Cell Biology. *Chem Rev.*, *109*(10), 4921–4960.

Katie J. Grayson & Ross Anderson, J. L. (2018). Designed for life: biocompatible de novo designed proteins and components. *J R Soc Interface.*, *15*(145), 20180472.

Kei Fukuhara, Shusaku Nagano, Mitsuo Hara & Takahiro Seki. (2014). Free-surface molecular command systems for photoalignment of liquid crystalline materials. *Nat Commun.*, *5*, 3320.

Kirsten Rasmussen, Hubert Rauscher, Agnieszka Mech, Juan Riego Sintes, Douglas Gilliland, Mar González, Peter Kearns, Kenneth Moss, Maaike Visser, Monique Groenewold & Eric A. J. Bleeker. (2018). Physico-chemical properties of manufactured nanomaterials - Characterisation and relevant methods. An outlook based on the OECD Testing Programme. *Regul Toxicol Pharmacol.*, *92*, 8–28.

Kjell De Vriese, Alex Costa, Tom Beeckman. & Steffen Vanneste. (2018). Pharmacological Strategies for Manipulating Plant Ca2+ Signalling. *Int J Mol Sci.*, *19*(5), 1506.

Kristian E. Dalle, Julien Warnan, Jane J. Leung, Bertrand Reuillard, Isabell S. Karmel & Erwin Reisner. (2019). Electro- and Solar-Driven Fuel Synthesis with First Row Transition Metal Complexes. *Chem Rev.*, *119*(4), 2752–2875.

Lara Donaldson, Stuart Meier & Christoph Gehring. (2016). The arabidopsis cyclic nucleotide interactome. *Cell Commun Signal.*, *14*, 10.

Larisa Zemskova, Andrei Egorin, Eduard Tokar, Vladimir Ivanov & Svetlana Bratskaya. (2018). New Chitosan/Iron Oxide Composites: Fabrication and Application for Removal of Sr^{2+} Radionuclide from Aqueous Solutions. *Biomimetics (Basel).*, *3*(4), 39.

Lidong Zhang, Haoran Liang, Jolly Jacob & Panče Naumov (2015). Photogated humidity-driven motility. *Nat Commun.*, *6*, 7429. 6, 7862.

LijiSobhana S. Sobhanadhas, Lokesh Kesavan & Pedro Fardim. (2018). Topochemical Engineering of Cellulose-Based Functional Materials. *Langmuir.*, *34*(34), 9857–9878.

Lijuan Luo, Xueying Lai, Baowei Chen, Li Lin, Ling Fang, Nora F. Y. Tam & Tiangang Luan. (2015). Chlorophyll catalyse the phototransformation of carcinogenic benzo[a]pyrene in water. *Sci Rep.*, *5*, 12776.

Ling-Li Min, Lu-Bin Zhong, Yu-Ming Zheng, Qing Liu, Zhi-Huan Yuan & Li-Ming Yang. (2016). Functionalized chitosan electrospun nanofiber for effective removal of trace arsenate from water. *Sci Rep.*, *6*, 32480.

Liping Song, Lei Zhang, Youju Huang, Liming Chen, Ganggang Zhang, Zheyu Shen, Jiawei Zhang, Zhidong Xiao & Tao Chen. (2017). Amplifying the signal of localized surface plasmon resonance sensing for the sensitive detection of Escherichia coli O157:H7. *Sci Rep.*, *7*, 3288.

Livy Laysandra, Meri Winda Masnona Kartika Sari, Felycia Edi Soetaredjo, Kuncoro Foe, Jindrayani Nyoo Putro, Alfin Kurniawan, Yi-Hsu Ju & Suryadi Ismadji. (2017). Adsorption and photocatalytic performance of bentonite-titanium dioxide composites for methylene blue and rhodamine B decoloration. *Heliyon.*, *3*(12), e00488.

Luca De Vico, André Anda, Vladimir Al. Osipov, Anders Ø. Madsen & Thorsten Hansen. (2018). Macrocycle ring deformation as the secondary design principle for light-harvesting complexes. *Proc Natl Acad Sci U S A.*, *115*(39), E9051–E9057.

Łukasz Drewniak, Robert Stasiuk, Witold Uhrynowski & Aleksandra Sklodowska. (2015). Shewanella sp. O23S as a Driving Agent of a System Utilizing Dissimilatory Arsenate-Reducing Bacteria Responsible for Self-Cleaning of Water Contaminated with Arsenic. *Int J Mol Sci.*, *16*(7), 14409–14427.

Lünsdorf, H., Brümmer, I., Timmis, K. N. & Wagner-Döbler, I. (1997). Metal selectivity of *in situ* microcolonies in biofilms of the Elbe river. *J Bacteriol.*, *179*(1), 31–40.

Maan Hayyan, Ali Abo-Hamad, Mohammed AbdulHakim AlSaadi & Mohd Ali Hashim. (2015). Functionalization of graphene using deep eutectic solvents. *Nanoscale Res Lett.*, *10*, 324.

Maciej Janek, Aleksandra Radtke, Tadeusz M. Muzioł, Maria Jerzykiewicz & Piotr Piszczek. (2018). Tetranuclear Oxo-Titanium Clusters with Different Carboxylate Aromatic Ligands: Optical Properties, DFT Calculations, and Photoactivity. *Materials (Basel).*, *11*(9), 1661.

Mahdi Karimi, Parham Sahandi Zangabad, Soodeh Baghaee-Ravari, Mehdi Ghazadeh, Hamid Mirshekari & Michael R. Hamblin. (2017). Smart Nanostructures for Cargo Delivery: Uncaging and Activating by Light. *J Am Chem Soc.*, *139*(13), 4584–4610.

Malatesta, M., Zancanaro, C., Costanzo, M., Cisterna, B. & Pellicciari, C. (2013). Simultaneous Ultrastructural Analysis of Fluorochrome-Photoconverted Diaminobenzidine and Gold Immunolabelling in Cultured Cells. *Eur J Histochem.*, *57*(3), e26.

Marco Felici, Pablo Contreras-Carballada, Jan M. M. Smits, Roeland J. M. Nolte, René M. Williams, Luisa De Cola & Martin C. Feiters (2010). Cationic Heteroleptic Cyclometalated IridiumIII Complexes Containing Phenyl-Triazole and Triazole-Pyridine Clicked Ligands. *Molecules.*, *15*(3), 2039–2059.

Maria Laura Tummino, Maria Luisa Testa, Mery Malandrino, Roberta Gamberini, Alessandra Bianco Prevot, Giuliana Magnacca & Enzo

Laurenti. (2019). Green Waste-Derived Substances Immobilized on SBA-15 Silica: Surface Properties, Adsorbing and Photosensitizing Activities towards Organic and Inorganic Substrates. *Nanomaterials (Basel).*, *9*(2), 162.

Mariana Q. Mesquita, Cristina J. Dias, Maria G. P. M. S. Neves, Adelaide Almeida, M. & Amparo F. Faustino. (2018). Revisiting Current Photoactive Materials for Antimicrobial Photodynamic Therapy. *Molecules.*, *23*(10), 2424.

Marzieh Azizi, Hedayatollah Ghourchian, Fatemeh Yazdian, Fariba Dashtestani & Hojjat Alizadeh Zeinabad. (2017). Cytotoxic effect of albumin coated copper nanoparticle on human breast cancer cells of MDA-MB 231. *PLoS One.*, *12*(11), e0188639.

Matthew S. Brown, Brandon Ashley & Ahyeon Koh. (2018). Wearable Technology for Chronic Wound Monitoring: Current Dressings, Advancements, and Future Prospects. *Front Bioeng Biotechnol.*, *6*, 47.

Mohamed Dawod, Natalie E. Arvin & Robert T. Kennedy. (2018). *Recent advances in protein analysis by capillary and microchip electrophoresis.*, *142*(11), 1847–1866.

Mohammed Kadhom, Weiming Hu & Baolin Deng. (2017). Thin Film Nanocomposite Membrane Filled with Metal-Organic Frameworks UiO-66 and MIL-125 Nanoparticles for Water Desalination. *Membranes (Basel).*, *7*(2), 31.

Muhammad Faizan, Niaz Muhammad, Kifayat Ullah Khan Niazi, Yongxia Hu, Yanyan Wang, Ya Wu, Huaming Sun, Ruixia Liu, Wensheng Dong, Weiqiang Zhang & Ziwei Gao. (2019). CO-Releasing Materials: An Emphasis on Therapeutic Implications, as Release and Subsequent Cytotoxicity Are the Part of Therapy. *Materials (Basel).*, *12*(10), 1643.

Natallia Makarava, Alexander Parfenov & Ilia V. Baskakov. (2005). Water-Soluble Hybrid Nanoclusters with Extra Bright and Photostable Emissions: A New Tool for Biological Imaging. *Biophys J.*, *89*(1), 572–580.

Navid B. Saleh, A. R. M. Nabiul Afrooz, Joseph H. Bisesi, Jr., Nirupam Aich, Jaime Plazas-Tuttle & Tara Sabo-Attwood. (2014). Emergent

Properties and Toxicological Considerations for Nanohybrid Materials in Aquatic Systems. *Nanomaterials (Basel).*, *4*(2), 372–407.

Nhung Thi Tuyet Le, Hirofumi Nagata, Mutsumi Aihara, Akira Takahashi, Toshihiro Okamoto, Takaaki Shimohata, Kazuaki Mawatari, Yhosuke Kinouchi, Masatake Akutagawa & Masanobu Haraguchi. (2011). Additional Effects of Silver Nanoparticles on Bactericidal Efficiency Depend on Calcination Temperature and Dip-Coating Speed. *Appl Environ Microbiol.*, *77*(16), 5629–5634.

Nina Kasyanenko, Zhang Qiushi, Vladimir Bakulev, Mikhail Osolodkov, Petr Sokolov & Viktor Demidov. (2017). DNA Binding with Acetate Bis(1,10-phenanthroline)silver(I) Monohydrate in a Solution and Metallization of Formed Structures. *Polymers (Basel).*, *9*(6), 211.

Nthabeleng Hlapisi, Tshwafo E. Motaung, Linda Z. Linganiso, Oluwatobi S. Oluwafemi & Sandile P. Songca. (2019). Encapsulation of Gold Nanorods with Porphyrins for the Potential Treatment of Cancer and Bacterial Diseases: A Critical Review. *Bioinorg Chem Appl.*, 2019, 7147128.

Olga E. Petrova & Karin Sauer. (2012). PAS Domain Residues and Prosthetic Group Involved in BdlA-Dependent Dispersion Response by Pseudomonas aeruginosa Biofilms. *J Bacteriol.*, *194*(21), 5817–5828.

Papri Sutar, Venkata M. Suresh, Kolleboyina Jayaramulu, Arpan Hazra & Tapas Kumar Maji. (2018). Binder driven self-assembly of metal-organic cubes towards functional hydrogels. *Nat Commun.*, *9*, 3587.

Patrick J. Snyder, Dennis R. LaJeunesse, Pramod Reddy, Ronny Kirste, Ramon Collazo & Albena Ivanisevic. (2018). Bioelectronics Communication: Encoding Yeast Regulatory Responses Using Nanostructured Gallium Nitride Thin Films. *Nanoscale.*, *10*(24), 11506–11516.

Patrick K. Jjemba, William Johnson, Zia Bukhari & Mark W. LeChevallier (2015). Occurrence and Control of Legionella in Recycled Water Systems. *Pathogens.*, *4*(3), 470–502.

Pedro Jorge, Manuel António Martins, Tito Trindade, José Luís Santos & Faramarz Farahi. (2007). Optical Fiber Sensing Using Quantum Dots. *Sensors (Basel).*, *7*(12), 3489–3534.

Peter S. Toth, Quentin M. Ramasse, Matěj Velický. & Robert A. W. Dryfe. (2015). Functionalization of graphene at the organic/water interface. *Chem Sci.*, *6*(2), 1316–1323.

Petra Jackson, Nicklas Raun Jacobsen, Anders Baun, Renie Birkedal, Dana Kühnel, Keld Alstrup Jensen, Ulla Vogel & Håkan Wallin. (2013). Bioaccumulation and ecotoxicity of carbon nanotubes. *Chem Cent J.*, 7, 154.

Pradeep Ramiah Rajasekaran, Chuanhong Zhou, Mallika Dasari, Kay-Obbe Voss, Christina Trautmann & Punit Kohli. (2017). Polymeric lithography editor: Editing lithographic errors with nanoporous polymeric probes. *Sci Adv.*, *3*(6), e1602071.

Purificación Cabello, Víctor M Luque-Almagro, Alfonso Olaya-Abril, Lara P Sáez, Conrado Moreno-Vivián & Dolores Roldán, M. (2018). Assimilation of cyanide and cyano-derivatives by Pseudomonas pseudoalcaligenes CECT5344: from omic approaches to biotechnological applications. *FEMS Microbiol Lett.*, *365*(6), fny032.

Purinergic Signal. (2014). Abstracts from Purines 2014, *an International Conference on Nucleotides, Nucleosides and Nucleobases*, held in Bonn, Germany, from July 23–27, *10*(4), 657–854.

Qiang Gao, Jun Luo, Xingyue Wang, Chunxia Gao & Mingqiao Ge. (2015). Novel hollow α-Fe2O3 nanofibers via electrospinning for dye adsorption. *Nanoscale Res Lett.*, *10*, 176.

Qingning Bian, Shasha Gao, Jilin Zhou, Jian Qin, Allen Taylor, Elizabeth J. Johnson, Guangwen Tang, Janet R. Sparrow, Dennis Gierhart & Fu Shang. (2012). Lutein and zeaxanthin supplementation reduces photo-oxidative damage and modulates the expression of inflammation-related genes in retinal pigment epithelial cells. *Free Radic Biol Med.*, *53*(6), 1298–1307.

Qiuxia Han, Bo Qi, Weimin Ren, Cheng He, Jingyang Niu & Chunying Duan. (2015). Polyoxometalate-based homochiral metal-organic frameworks for tandem asymmetric transformation of cyclic carbonates from olefins. *Nat Commun.*, *6*, 10007.

Rachel J. Service, Warwick Hillier & Richard J. Debus. (2011). Evidence from FTIR Difference Spectroscopy for an Extensive Network of Hydrogen Bonds near the Oxygen-Evolving Mn4Ca cluster of Photosystem II Involving D1-Glu65, D2-Glu312, and D1-Glu329. *Biochemistry.*, *49*(31), 6655–6669.

Rashmi A. Agarwal, Neeraj K. Gupta, Rajan Singh, Shivansh Nigam & Bushra Ateeq. (2017). Ag/AgO Nanoparticles Grown via Time Dependent Double Mechanism in a 2D Layered Ni-PCP and Their Antibacterial Efficacy. *Sci Rep.*, *7*, 44852.

Raymond R. Suskind. (1977). Environment and the skin. Environ Health Perspect.

Reddi, E., Zhou, C., Biolo, R., Menegaldo, E. & Jori, G. (1990). Liposome- or LDL-administered Zn (II)-phthalocyanine as a photodynamic agent for tumours. I. Pharmacokinetic properties and phototherapeutic efficiency. *Br J Cancer.*, *61*(3), 407–411.

Reheman, A., Tursun, Y., Dilinuer, T., Halidan, M., Kadeer, K. & Abulizi, A. (2018). Facile One-Step Sonochemical Synthesis and Photocatalytic Properties of Graphene/Ag3PO4 Quantum Dots Composites. *Nanoscale Res Lett.*, *13*, 70.

Renlan Liu, Xiaoying Zhu & Baoliang Chen. (2017). A New Insight of Graphene oxide-Fe(III) Complex Photochemical Behaviors under Visible Light Irradiation. *Sci Rep.*, *7*, 40711.

Robbin R. Vernooij, Dr. Tanmaya Joshi, Dr. Michael D. Horbury, Assoc., Prof. Bim Graham, Assoc., Prof. Ekaterina I. Izgorodina, Prof. Vasilios G. Stavros, Prof. Peter J. Sadler, Prof. Leone Spiccia & Assoc., Prof. Bayden R. Wood. (2018). Spectroscopic Studies on Photoinduced Reactions of the Anticancer Prodrug, trans,trans,trans-[Pt(N3)2(OH)2(py)2]. *Chemistry.*, *24*(22), 5790–5803.

Roger J. Kutta, Samantha J. O. Hardman, Linus O. Johannissen, Bruno Bellina, Hanan L. Messiha, Juan Manuel Ortiz-Guerrero, Montserrat Elías-Arnanz, S. Padmanabhan, Perdita Barran & Nigel S. Scrutton.

(2015). Alex R. Jones. The photochemical mechanism of a B12-dependent photoreceptor protein. *Nat Commun.*, *6*, 7907.

Roger J. Narayan, Shashishekar P. Adiga, Michael J. Pellin, Larry A. Curtiss, Alexander J. Hryn, Shane Stafslien, Bret Chisholm, Chun-Che Shih, Chun-Ming Shih, Shing-Jong Lin, Yea-Yang Su, Chunming Jin, Junping Zhang, Nancy A. Monteiro-Riviere & Jeffrey W. Elam. (2010). *Atomic layer deposition-based functionalization of materials for medical and environmental health applications Philos Trans A Math Phys Eng Sci.*, *368*(1917), 2033–2064.

Roland Krivanek, Holger Dau & Michael Haumann. (2008). Enthalpy Changes during Photosynthetic Water Oxidation Tracked by Time-Resolved Calorimetry Using a Photothermal Beam Deflection Technique. *Biophys J.*, *94*(5), 1890–1903.

Ronald B. Walter, Dylan J. Walter, William T. Boswell, Kaela L. Caballero, Mikki Boswell, Yuan Lu, Jordan Chang & Markita G. Savage. (2015). *Exposure to Fluorescent Light Triggers Down Regulation of Genes Involved with Mitotic Progression in Xiphophorus Skin Comp Biochem Physiol C Toxicol Pharmacol.*, *178*, 93–103.

Rui Gao & Dongpeng Yan. (2017). Layered host–guest long-afterglow ultrathin nanosheets: high-efficiency phosphorescence energy transfer at 2D confined interface. *Chem Sci.*, *8*(1), 590–599.

Rui Liu, Mengran Liu, Don Hood, Chih-Yuan Chen, Christopher J. MacNevin, Dewey Holten & Jonathan S. Lindsey. (2018). Chlorophyll-Inspired Red-Region Fluorophores: Building Block Synthesis and Studies in Aqueous Media. *Molecules.*, *23*(1), 130.

Runfa Li, Yonghai Feng, Guoqing Pan & Lei Liu. (2019). Advances in Molecularly Imprinting Technology for Bioanalytical Applications. *Sensors (Basel).*, *19*(1), 177.

Sami Rtimi, Stefanos Giannakis & Cesar Pulgarin. (2017). Self-Sterilizing Sputtered Films for Applications in Hospital Facilities. *Molecules.*, *22*(7), 1074.

Samuel Clark Ligon, Robert Liska, Jürgen Stampfl, Matthias Gurr & Rolf Mülhaupt. (2017). Polymers for 3D Printing and Customized Additive Manufacturing. *Chem Rev.*, *117*(15), 10212–10290.

Saner Poplata, Andreas Tröster, You-Quan Zou & Thorsten Bach. (2016). Recent Advances in the Synthesis of Cyclobutanes by Olefin [2 + 2] Photocycloaddition Reactions. *Chem Rev.*, *116*(17), 9748–9815.

Sanjay M. Prakadan, Alex K. Shalek & David A. Weitz. (2017). Scaling by shrinking: empowering single-cell 'omics' with microfluidic devices. *Nat Rev Genet.*, *18*(6), 345–361.

Sanjib Bhattacharyya, Rachel A. Kudgus, Resham Bhattacharya & Priyabrata Mukherjee. (2011). Inorganic Nanoparticles in Cancer Therapy. *Pharm Res.*, *28*(2), 237–259.

Schweimer, K., Hoffmann, S., Wastl, J., Maier, U. G., Rösch, P. & Sticht, H. (2000). Solution structure of a zinc substituted eukaryotic rubredoxin from the cryptomonad alga Guillardia theta. *Protein Sci.*, *9*(8), 1474–1486.

Sergii Golovynskyi, Luca Seravalli, Oleksandr Datsenko, Oleksii Kozak, Serhiy V. Kondratenko, Giovanna Trevisi, Paola Frigeri, Enos Gombia, Sergii R. Lavoryk, Iuliia Golovynska, Tymish Y. Ohulchanskyy & Junle Qu. (2017). Bipolar Effects in Photovoltage of Metamorphic InAs/InGaAs/GaAs Quantum Dot Heterostructures: Characterization and Design Solutions for Light-Sensitive Devices. *Nanoscale Res Lett.*, *12*, 559.

Shaik Anwar Ahamed Nabeela Nasreen, Subramanian Sundarrajan, Syed Abdulrahim Syed Nizar & Seeram Ramakrishna. (2019). Nanomaterials: Solutions to Water-Concomitant Challenges. *Membranes (Basel).*, *9*(3), 40.

Shenghua Jiang & Hor-Gil Hur. (2013). Effects of the Anaerobic Respiration of Shewanella oneidensis MR-1 on the Stability of Extracellular U(VI) Nanofibers. *Microbes Environ.*, *28*(3), 312–315.

Soad Z. Al Sheheri, Zahra M. Al-Amshany, Qana A. Al Sulami, Nada Y. Tashkandi, Mahmoud A. Hussein & Reda M. El-Shishtawy. (2019). The preparation of carbon nanofillers and their role on the performance of variable polymer nanocomposites. *Des Monomers Polym.*, *22*(1), 8–53.

Sobia Ashraf, Asima Siddiqa, Shabnam Shahida & Sara Qaisar. (2019). *Titanium-based nanocomposite materials for arsenic removal from* water: A review *Heliyon.*, *5*(5), e01577. Correction in: *Heliyon.*, *5*(6), e01889.

Sonia Pérez-Rentero, Santiago Grijalvo, Rubén Ferreira & Ramon Eritja. (2012). Synthesis of Oligonucleotides Carrying Thiol Groups Using a Simple Reagent Derived from Threoninol. *Molecules.*, *17*(9), 10026–10045.

Sooim Shin, Moonsung Choi, Heather R. Williamson & Victor L. Davidson. (2014). A simple method to engineer a protein-derived redox cofactor for catalysis *Biochim Biophys Acta.*, *1837*(10), 1595–1601.

Soraya Taabache & Annabelle Bertin. (2017). Vesicles from Amphiphilic Dumbbells and Janus Dendrimers: Bioinspired Self-Assembled Structures for Biomedical Applications. *Polymers (Basel).*, *9*(7), 280.

Starukh, G. (2017). Photocatalytically Enhanced Cationic Dye Removal with Zn-Al Layered Double Hydroxides. *Nanoscale Res Lett.*, *12*, 391.

Stefanelli, M., Mancini, M., Raggio, M., Nardis, S., Fronczeck, F. R., McCandless, G. T., Smith, K. M. & Paolesse, R. (2014). 3-NO2-5,10,15-triarylcorrolato-Cu as a versatile platform for synthesis of novel 3-functionalized corrole derivatives. *Org Biomol Chem.*, *12*(32), 6200–6207.

Stephan Mändl. (2009). Increased Biocompatibility and Bioactivity after Energetic PVD Surface Treatments. *Materials (Basel).*, *2*(3), 1341–1387.

Stephanie J. Dancer. (2014). Controlling Hospital-Acquired Infection: Focus on the Role of the Environment and New Technologies for Decontamination. *Clin Microbiol Rev.*, *27*(4), 665–690.

Steven Y. Reece, Daniel A. Lutterman, Mohammad R. Seyedsayamdost, JoAnne Stubbe & Daniel G. Nocera. (2009). Re(bpy)(CO)3CN as a Probe of Conformational Flexibility in a Photochemical Ribonucleotide Reductase. *Biochemistry.*, *48*(25), 5832–5838.

Suman B. Mondal, Shengkui Gao, Nan Zhu, Gail P. Sudlow, Kexian Liang, Avik Som, Walter J. Akers, Ryan C. Fields, Julie Margenthaler, Rongguang Liang, Viktor Gruev & Samuel Achilefu. (2015). Binocular Goggle Augmented Imaging and Navigation System provides real-time fluorescence image guidance for tumor resection and sentinel lymph node mapping. *Sci Rep.*, *5*, 12117.

Sunaina Singh, Amit Aggarwal, N. V. S. Dinesh K. Bhupathiraju, Gianluca Arianna, Kirran Tiwari & Charles Michael Drain. (2015). Glycosylated Porphyrins, Phthalocyanines, and Other Porphyrinoids for Diagnostics and Therapeutics. *Chem Rev.*, *115*(18), 10261–10306.

Szymon Kugler, Paula Ossowicz, Kornelia Malarczyk-Matusiak & Ewa Wierzbicka. (2019). Advances in Rosin-Based Chemicals: The Latest Recipes, Applications and Future Trends. *Molecules.*, *24*(9), 1651.

Tae-Yang Kim, Min Gyu Kim, Ji-Hoon Lee & Hor-Gil Hur. (2018). Biosynthesis of Nanomaterials by Shewanella Species for Application in Lithium Ion Batteries. *Front Microbiol.*, *9*, 2817.

Teresa J. Bandosz & Conchi O. Ania. (2018). Origin and Perspectives of the Photochemical Activity of Nanoporous Carbons. *Adv Sci (Weinh).*, *5*(9), 1800293.

Thomas Riedel, Laura Gómez-Consarnau, Jürgen Tomasch, Madeleine Martin, Michael Jarek, José M. González, Stefan Spring, Meike Rohlfs, Thorsten Brinkhoff, Heribert Cypionka, Markus Göker, Anne Fiebig, Johannes Klein, Alexander Goesmann, Jed A. Fuhrman & Irene Wagner-Döbler. (2013). Genomics and Physiology of a Marine Flavobacterium Encoding a Proteorhodopsin and a Xanthorhodopsin-Like Protein. *PLoS One.*, *8*(3), e57487.

Thomson, N. H., Smith, B. L., Almqvist, N., Schmitt, L., Kashlev, M., Kool, E. T. & Hansma, P. K. (1999). Oriented, active Escherichia coli RNA polymerase: an atomic force microscope study. *Biophys J.*, *76*(2), 1024–1033.

Till Rudack, Sarah Jenrich, Sven Brucker, Ingrid R. Vetter, Klaus Gerwert & Carsten Kötting. (2015). Catalysis of GTP Hydrolysis by Small GTPases at Atomic Detail by Integration of X-ray Crystallography, Experimental, and Theoretical IR Spectroscopy. *J Biol Chem.*, *290*(40), 24079–24090.

Ukrae Cho, Daniel P. Riordan, Paulina Ciepla, Kiranmai S. Kocherlakota, James K. Chen & Pehr B. Harbury. (2018). Ultrasensitive optical imaging with lanthanide lumiphores. *Nat Chem Biol.*, *14*(1), 15–21.

Vassiliev, I. R., Kjaer, B., Schorner, G. L., Scheller, H. V. & Golbeck, J. H. (2001). Photoinduced transient absorbance spectra of P840/P840(+) and the FMO protein in reaction centers of Chlorobium vibrioforme. *Biophys J.*, *81*(1), 382–393.

Victoria Klang, Claudia Valenta & Nadejda B. Matsko. (2013). Analytical Electron Microscopy for Characterization of Fluid or Semi-Solid Multiphase Systems Containing Nanoparticulate Material. *Pharmaceutics.*, *5*(1), 115–126.

Vignesh Nayak, Mannekote Shivanna Jyothi, Prof. R Geetha Balakrishna, Dr. Mahesh Padaki & Prof. Ahmad Fauzi Ismail. (2015). Preparation and Characterization of Chitosan Thin Films on Mixed-Matrix Membranes for Complete Removal of Chromium. *Chemistry Open.*, *4*(3), 278–287.

Wataru Nishima, Wataru Mizukami, Yoshiki Tanaka, Ryuichiro Ishitani, Osamu Nureki & Yuji Sugita (2016). Mechanisms for Two-Step Proton Transfer Reactions in the Outward-Facing Form of MATE Transporter. *Biophys J.*, *110*(6), 1346–1354.

Wei Luo, Xingfeng Wang, Colin Meyers, Nick Wannenmacher, Weekit Sirisaksoontorn, Michael M. Lerner & Xiulei Ji. (2013). Efficient Fabrication of Nanoporous Si and Si/Ge Enabled by a Heat Scavenger in Magnesiothermic Reactions. *Sci Rep.*, *3*, 2222.

Weiwei Han, Zhen Li, Yang Li, Xiaobin Fan, Fengbao Zhang, Guoliang Zhang & Wenchao Peng. (2017). The Promoting Role of Different Carbon Allotropes Cocatalysts for Semiconductors in Photocatalytic Energy Generation and Pollutants Degradation. *Front Chem.*, *5*, 84.

Wenlong Xiang, Yueping Zhang, Hongfei Lin & Chang-jun Liu. (2017). Nanoparticle/Metal–Organic Framework Composites for Catalytic Applications: Current Status and Perspective. *Molecules.*, *22*(12), 2103.

Wilfred K. Fullagar, Jens Uhlig, Ujjwal Mandal, Dharmalingam Kurunthu, Amal El Nahhas, Hideyuki Tatsuno, Alireza Honarfar, Fredrik Parnefjord Gustafsson, Villy Sundström, Mikko R. J. Palosaari, Kimmo M. Kinnunen, Ilari J. Maasilta, Luis Miaja-Avila, Galen C. O'Neil, Young Il Joe, Daniel S. Swetz & Joel N. Ullom. (2017).

Beating Darwin-Bragg losses in lab-based ultrafast x-ray experiments. *Struct Dyn.*, *4*(4), 044011.

Wouter Maijenburg, Eddy J. B. Rodijk, Michiel G. Maas & Johan E. ten Elshof. (2014). Preparation and Use of Photocatalytically Active Segmented Ag|ZnO and Coaxial TiO2-Ag Nanowires Made by Templated Electrodeposition. *J Vis Exp.*, (87), 51547.

Xiaoling Yang, Wei Chen, Jianfei Huang, Ying Zhou, Yihua Zhu & Chunzhong Li. (2015). Rapid degradation of methylene blue in a novel heterogeneous Fe3O4@rGO@TiO2-catalyzed photo-Fenton system. *Sci Rep.*, *5*, 10632.

Xiaoqin Zhou, Zifu Li, Tianlong Zheng, Yichang Yan, Pengyu Li, Emmanuel Alepu Odey, Heinz Peter Mang. & Sayed Mohammad Nazim Uddin. (2018). Review of global sanitation development. *Environ Int.*, *120*, 246–261.

Xiaoyun Qiu. & Shuwen Hu. (2013). "Smart" Materials Based on Cellulose: A Review of the Preparations, Properties, and Applications. *Materials (Basel).*, *6*(3), 738–781.

Xing Liu, Limei Li, Meijuan Li, Liangchen Su, Siman Lian, Baihong Zhang, Xiaoyun Li, Kui Ge & Ling Li. (2018). AhGLK1 affects chlorophyll biosynthesis and photosynthesis in peanut leaves during recovery from drought. *Sci Rep.*, *8*, 2250.

Xuenan Feng, Chenxi Liu, Xiqian Wang, Yuying Jiang, Gengxiang Yang, Rong Wang, Kaishun Zheng, Weixiao Zhang, Tianyu Wang & Jianzhuang Jiang. (2019). Functional Supramolecular Gels Based on the Hierarchical Assembly of Porphyrins and Phthalocyanines. *Front Chem.*, *7*, 336.

Xue-Xue Liang, Nan Wang, You-Le Qu, Li-Ye Yang, Yang-Guang Wang & Xiao-Kun Ouyang. (2018). Facile Preparation of Metal-Organic Framework (MIL-125)/Chitosan Beads for Adsorption of Pb(II) from Aqueous Solutions. *Molecules.*, *23*(7), 1524.

Ya Tuo, Guangfei Liu, Bin Dong, Jiti Zhou, Aijie Wang, Jing Wang, Ruofei Jin, Hong Lv, Zeou Dou & Wenyu Huang. (2015). Microbial synthesis of Pd/Fe3O4, Au/Fe3O4 and PdAu/Fe3O4 nanocomposites for catalytic reduction of nitroaromatic compounds. *Sci Rep.*, *5*, 13515.

Yanan Zou, Yue Zhang, Yongming Hu & Haoshuang Gu. (2018). Ultraviolet Detectors Based on Wide Bandgap Semiconductor Nanowire: A Review. *Sensors (Basel)*, *18*(7), 2072.

Yanke Wei, Lefu Mei, Rui Li, Meng Liu, Guocheng Lv, Jianle Weng, Libing Liao, Zhaohui Li & Lin Lu. (2018). Fabrication of an AMC/MMT Fluorescence Composite for its Detection of Cr(VI) in Water. *Front Chem.*, *6*, 367.

Yanpei Song, Qi Sun, Briana Aguila & Shengqian Ma. (2019). Opportunities of Covalent Organic Frameworks for Advanced Applications. *Adv Sci (Weinh).*, *6*(2), 1801410.

Ya-Ping Sun, Ping Wang, Zhuomin Lu, Fan Yang, Mohammed J. Meziani, Gregory E. LeCroy, Yun Liu & Haijun Qian. (2015). Host-Guest Carbon Dots for Enhanced Optical Properties and Beyond. *Sci Rep.*, *5*, 12354.

Yaron Paz. (2011). Self-assembled monolayers and titanium dioxide: From surface patterning to potential applications. *Beilstein J Nanotechnol.*, *2*, 845–861.

Yihang Chu, Chunqi Qian, Premjeet Chahal & Changyong Cao. (2019). Printed Diodes: Materials Processing, Fabrication, and Applications. *Adv Sci (Weinh).*, *6*(6), 1801653.

Yilong Han, Lidietta Giorno & Annarosa Gugliuzza. (2017). Photoactive Gel for Assisted Cleaning during Olive Mill Wastewater Membrane Microfiltration. *Membranes (Basel).*, *7*(4), 66.

Ying-Ying Huang, Sulbha K. Sharma, Tianhong Dai, Hoon Chung, Anastasia Yaroslavsky, Maria Garcia-Diaz, Julie Chang, Long Y. Chiang & Michael R. Hamblin. (2012). Can nanotechnology potentiate photodynamic therapy? *Nanotechnol Rev.*, *1*(2), 111–146.

Yohei Takashima, Virginia Martínez Martínez, Shuhei Furukawa, Mio Kondo, Satoru Shimomura, Hiromitsu Uehara, Masashi Nakahama, Kunihisa Sugimoto & Susumu Kitagawa. (2011). Molecular decoding using luminescence from an entangled porous framework. *Nat Commun.*, *2*, 168.

York R. Smith, Rupashree S. Ray, Krista Carlson, Biplab Sarma & Mano Misra. (2013). Self-Ordered Titanium Dioxide Nanotube Arrays:

Anodic Synthesis and Their Photo/Electro-Catalytic Applications. *Materials (Basel).*, *6*(7), 2892–2957.

Yun Zheng, Zihao Yu, Feng Lin, Fangsong Guo, Khalid A. Alamry, Layla A. Taib, Abdullah M. Asiri & Xinchen Wang. (2017). Sulfur-Doped Carbon Nitride Polymers for Photocatalytic Degradation of Organic Pollutant and Reduction of Cr(VI). *Molecules.*, *22*(4), 572.

Yvonne N. Tallini, Masamichi Ohkura, Bum-Rak Choi, Guangju Ji, Keiji Imoto, Robert Doran, Jane Lee, Patricia Plan, Jason Wilson, Hong-Bo Xin, Atsushi Sanbe, James Gulick, John Mathai, Jeffrey Robbins, Guy Salama, Junichi Nakai & Michael I. Kotlikoff. (2006). Imaging cellular signals in the heart *in vivo*: Cardiac expression of the high-signal Ca2+ indicator GCaMP2. *Proc Natl Acad Sci U S A.*, *103*(12), 4753–4758.

Zhengqiang Xia, Cheng He, Xiaoge Wang & Chunying Duan. (2017). Modifying electron transfer between photoredox and organocatalytic units via framework interpenetration for β-carbonyl functionalization. *Nat Commun.*, *8*, 361.

Zhong Wu, Lin Li, Jun-min Yan & Xin-bo Zhang. (2017). Materials Design and System Construction for Conventional and New-Concept Supercapacitors. *Adv Sci (Weinh).*, *4*(6), 1600382.

Zhuang Liu, Scott Tabakman, Kevin Welsher & Hongjie Dai. (2009). Carbon Nanotubes in Biology and Medicine: *In vitro* and *in vivo* Detection, Imaging and Drug Delivery. *Nano Res.*, *2*(2), 85–120.

In: Environmental Science of Heavy Metals ISBN: 978-1-53617-831-9
Editor: Dorota Bartusik-Aebisher © 2020 Nova Science Publishers, Inc.

Chapter 3

BIOMASS-BASED ABSORBENTS FOR HEAVY METAL REMOVAL

Tomasz Kubrak, Rafał Podgórski, David Aebisher[*],
Sabina Galiniak and Dorota Bartusik-Aebisher
Faculty of Medicine, University of Rzeszow, Poland

ABSTRACT

Continuing development of civilization and urbanization has resulted in an enormous amount of heavy metals being released into the environment. Heavy metals are highly toxic, and pose a significant threat not only to human and animal health, but also to the entire natural environment. In order to eliminate pollutants from the environment, numerous techniques have been developed using biological materials, i.e., bio-sorbents. The most important features of bio-sorbents are that they are often renewable, biodegradable and have low operating costs. Each biomass has the property of binding metal ions, however, depending on its type, the capacity to bind metals and bonding mechanisms differs. Bio-sorbents can be classified as biomass, e.g., plants (moss, leaves,

---

[*] Corresponding Author's Email: dbartusik-aebisher@ur.edu.pl.

trees), algae, bacteria, fungi or yeasts. Different biomass processing techniques can contribute to increasing the removal of metal ions. Research shows that various modifications of biomass significantly improve its ability to adsorb heavy metal ions.

A literature review in this chapter examines the possibilities of using and processing selected bio-sorbents for removal of heavy metal ions from polluted environments by characterizing biomass-based absorbents (e.g., aquatic biomass, terrestrial biomass, soil and mineral deposits and agricultural waste products) for heavy metal removal.

Keywords: heavy metals, adsorption, bio-sorption, biomass

INTRODUCTION

Recently, the whole developing world has seen an increase in industrial and urban activity which has contributed to increasing pollution of the surrounding environment with heavy metals. Numerous industrial plants release pollutants from wastewater (power plants, mining), and recycling and disposal is the main source of pollution (Herat and Agamuthu 2012; Olafisoye et al. 2013; Perkins et al. 2014; Wu et al. 2015) (Demirak et al. 2006) and mining (Archundia et al. 2017 Emissions from vehicles and other activities in urban areas also have a very large impact on environmental pollution (Pandey and Pandey 2009; Prasse et al. 2012).

The main sources of anthropogenic metals include combustion processes in power plants, industrial plants, road transport, production processes without combustion, metal ore mining, and waste utilization. As, Cd, Cr, Cu, Hg, Pb, Zn, Sb. Co, and Ni are considered to be particularly dangerous in the environment and are classified as heavy metals (Tóth et al. 2016). Heavy metals found in the surrounding human environment are particularly dangerous due to their toxic and carcinogenic nature. Their harm to human health has been repeatedly researched and documented (Jarup 2003; Tchounwou et al. 2012; Chakraborty et al. 2013; Schwartzbord et al. 2013; Jaishankar et al. 2014; Gleason et al. 2016; Masindi and Muedi 2018). Health problems related to long-term exposure to heavy metals in humans and animals cause disorders of the central nervous system (As), mental disabilities (Hg), diseases of the kidneys and liver (Hg, Cd), and skin damage (As) (Jaishankar et al. 2014). Heavy metals can be efficiently removed by "green" bio-sorbents, such as fungal biomasses, marine algae, agricultural waste and residues, yeasts, bacteria as well as biosorbents containing chitosan made from shellfish coatings.

Many agricultural wastes, such as fertilizers, bark or compost, contain large amounts of ligno-cellulosic substances that can be used for the removal of heavy metals.

TERRESTRIAL BIOMASS

Many species of trees, plants and aquatic animals exhibit lesser or greater degrees of bioremediation capacity. The first example is *Moringa oleifera*. This plant is the most widely cultivated species in the genus Moringa, and is a fast-growing, drought-resistant tree. It is widely cultivated for its use in herbal medicine (Bhattacharya et al. 2018). Research has shown that *Moringa oleifera* has the ability to remove heavy metals, enabling purification of drinking water and wastewater (Kalibbala et al. 2009; Shan et al. 2017) with an overall efficiency of 90% in the elimination of harmful elements from river and sewage (for Cd (II), Fe (II), Cr (III), Zn (II) and Cu (II) (Kansal and Kumari 2014; Shan et al. 2017). On the other hand, bioremediation of Pb (II) from wastewater was not demonstrated (Shan et al. 2017).

Another example of a plant effective in removing Zn (II), Cu (II) and Pb (II) are tree ferns. This type of wood fern grows in Taiwan and is usually sold for horticultural purposes, its fibers being used for pots due to its ability to absorb water for potted plants. Absorption abilities result from the fern's fiber construction from cellulose. Cellulose has a negative charge and strongly attracts positively charged metal cations (Ho et al. 2002). Srivastava et al. (2015) demonstrated the ability of the bark of *Lagerstroemia speciosa*, a tree originating in India, to absorb Cr (VI). *Cassia fistula* leaf extract has been shown to effectively remove chromium

(VI) (Ahmad et al. 2017), while *Camphora cinnamomum* species leaves absorb Pb (II) (Chen et al. 2010).

Many studies have focused on effective elimination of heavy metals using sawdust from numerous tree species. For example, maple sawdust is characterized by high efficiency of Cr (VI) removal (80%) (Yu et al., 2003). Beech sawdust removes Cu (II), Ni (II), Cd (II) and Zn (II). Similar research has shown that sawdust from poplar and linden trees also can aid in removal of Zn (II), Ni (II), Cd (II), Cu (II) and Mn(II) (Bozic et al. 2009). Moreover, in other work, Cr (III), Cu (II) and Pb (II) were removed using sawdust from poplar trees (Li et al. 2007),and Cu (II), Ni (II) and Zn (II) were effectively absorbed by teakwood sawdust (Shukla and Pai 2005). Shukla and Pai (2005) increased the adsorption capacity of sawdust from 40% to 70% by chemical modification.

Another absorbent classified as biomass that often has been tested for removal of heavy metals is lignin. Lignin is a basic components of wood in addition to cellulose and hemicellulose that is 20% by weight. Chemically, lignin is a polymer whose monomers are organic compounds that are derived from phenol alcohols. Polymeric phenols provides wood with compressive strength and aids in maintaining stiffness. It is suspected that lignin eliminates heavy metals due to the presence of functional groups within its structure. Heavy metals strongly react with carboxyl and phenol groups that comprise lignin, and are ideal metal adsorption sites (Guo et al. 2008).

Numerous studies demonstrate the effectiveness of using lignin as a heavy metal adsorbent. Guo et al. (2008) successfully used lignin, which is a waste product from the production of paper. It absorbs impurities from water in the form of heavy metals such as Pb (II), Cu (II), Zn (II) and Ni (II). Other studies indicate the effective use of lignin for removal of Cr(VI) metals (Liang et al. 2013) and Cr (III) (Wu et al. 2008). The phenomenon of bioremediation and use of lignin and other lignocelluloses in the removal of heavy metals has been the subject of several review articles recently (Ge and Li 2018; Neris et al. 2019).

AQUATIC BIOMASS

There are many examples of the use of water biomasses, such as algae and seaweedare characterized by high bioremediation ability being used to purify water containing heavy metals.

Algae are biomasses that naturally spreads in various places. Advantages of algae as bio-sorbents are high availability, low acquisition costs, and large sorption capacity and a relatively constant composition. Onyancha et al. (2008) used two species of algae bio-sorbents in their study: *Spirogyra condensata* and *Rhizoclonium hieroglyphicum*. They effectively removed Cr (III) from tannery effluent. Sari and Tuzen (2008) focused on the bio-sorption of total chromium from aqueous solution. They obtained over 90% efficiency of metal removal using a *Ceramium virgatum* (red algae) biomass.

Other studies have shown the effectiveness of Ni (II) and Cu (II) removal with the marine algae *Undaria pinnatifida* (Asian kelp). Cd (II), Ni (II) and Zn (II) were also removed using *Ascophyllum nodosum* (brown), *Chondrus crispus* (red), and *Codium vermilara* (green) algae as biosorbents. High metal removal efficiency was obtained using both single species and different species combinations (Romera et al. 2008). Hashim and Chu (2004) investigated the adsorption capacity of Cd (II) removal using species of seaweed. Brown seaweed *Sargassum baccularia* proved to be the most effective followed by red seaweed *Gracilaria salicornia* and least effective turned out to be green seaweed *Chaetomorpha linum*. In studies by Ajjabi and Chouba (2009), Cu(II) and Zn(II) was removed from wastewater using dried sea algae *Chaetomorpha linum*, Cu(II), Cd(II), and Pb(II) by dried green macroalga *Caulerpa lentillifera (Apiratikul and Pavasant 2006),* toxic chromium ions by *Ulva lactuca* (El-Sikaily et al. 2007) and Cu(II) with Pb(II) by *Cladophora fascicularis* (Deng et al. 2006). Other research showed that biomass from the freshwater macrophyte *Spirodela polyrhiza* could be used as an effective adsorbent for Cu (II), Mn (II) and Zn (II) ion removal from wastewaters (Meitei and Prasad 2014).

The use in water lilies and mangrove leaves has indicated a potential for chromium adsorption (Elangovan et al. 2008). Other aquatic weeds, such as hyacinth roots and neem leaves, have been found to remove Cu(II) (Singha and Das 2013).

Chemical modifications of biomasses are a commonly used method used to increase or reduce removal of heavy metals from the environment. The addition of 0.2M $CaCl_2$ to extract obtained from *Undaria pinnatifida* biomass increased the adsorption capacity of Cu (II) and Ni (II) removal (Chen et al. 2008). However, rinsing water weeds such as, *Cannomois Vvirgata* (reed mat), *Pistia stratiotes* (water lettuce), *Polygonum sagittatum* (arrow-leaved tear thumb), *Nelumbo nucifera* (lotus flower), *Colocasia esculenta* (green taro), *Nymphaea sp.* (Water lily flower), *Eichornia crassipes* (water hyacinth), and *Rhizophora mangle* (mangrove leaves) with 4N H_2SO_4 or NaOH solution caused a decrease in Cr (III) absorption capacity (Elangovan et al. 2008).

Research into increasing the efficiency of bio-sorption and bioaccumulation processes is ongoing. Different biomass processing techniques can contribute to increasing the removal of metal ions. The presented examples show that various types of biomass modification can significantly improve a biomass's ability to adsorb heavy metal ions.

AGRICULTURAL WASTE PRODUCTS

Agricultural waste is a rich source of effective adsorbents that can be used for heavy metal removal processes. One of the most common adsorbents is compost, a unique material that is effective in removing heavy metals, primarily Zn (II), Cu (II) and Pb (II) (Zhang 2011).

Produced in large quantities, especially in Asian countries, waste residues from rice production (rice bran, straw) are able to remove Cu (II) ions (Singha and Das 2013). In turn, rice pulp removed the largest amount of Pb (II) ions, Cu (II) and Hg (II), and Zn (II), Cd (II), Mn (II), Co (II) in smaller amountsand trace amounts of Ni (II) (Krishnani et al. 2008).

In many studies, chemical modifications of rice husks have resulted in increased assimilation capacity of Cr (VI) ions by lowering the pH below 3.0 (Georgieva et al. 2015), or in the presence of formaldehyde (Bansal et al. 2009). In contrast, the presence of phosphate in the environment of rice husks increased adsorption capacity towards Cd (II) ions (Ajmal et al. 2003).

Ground peanut shell was also tested for removal of heavy metals. A high adsorption of Pb (II) ions was obtained. In order to increase the efficiency of the process, reaction conditions of pH and temperature were changed (Taşar et al. 2014). Ahmad et al. (2017) used low pH values and obtained adsorption of Cr (VI) ions using nut shells. Peanut shells have been the subject of many scientific works in bioremediation as they are able to effectively remove Cr (III) and Cu (II) (Witek-Krowiak et al. 2011). Li et al. (2007) obtained adsorption of Cr (III), Cu (II) and Pb (II) and confirmed the removal of Cu (II) ions reported by Zhu et al. (2009). Subjecting peanut shells to chemical modification results in an increased adsorption capacity for Cu (II), Ni (II) and Zn (II) ions (Shukla and Pai 2005).

It was found that the use of other nut species also contributes to purification of, for example, heavy metal from water. Shells of cashew nuts significantly (85%) remove Cu (II) ions (Senthil Kumar et al. 2011). The same scientists continued their research and obtained adsorption of Ni (II) ions (Senthil Kumar et al. 2011). Researchers hypothesized that the adsorption capacity of nut shells is the result of surface features. Numerous ripples increase the surface area considerably and contributes to an increase in the number of active sites (Senthil Kumar et al. 2011, Senthil Kumar et al. 2011). Residues from pistachio shells were tested for removing Cr (VI) ions from contaminated samples. The high adsorptive capacity resulted from the diversity of functional groups on the adsorbent surface providing sites where Cr (VI) ions could be effectively attracted by electrostatic bonds (Moussavi and Barikbin 2010).

Pecan nuts shells with additional chemical modification such as addition of acid, and water vapor improved removal of Cu (II), Pb (II) and Zn (II) ions. Pb (II) ions were adsorbed in the largest amount, and lower adsorption of Cu (II) and Zn (II) in the acidic environment was noted (Bansode et al. 2003).

The use of a suitable dose of almond nut shells and a suitable pH resulted in an increase in Cr (VI) adsorption potential (Dakiky et al., 2002). Demirbas et al. 2009 successfully used hazelnut shells to remove Cu (II) ions.

Also, for removal of heavy metals from natural habitats, wastes from various citrus fruits were tested. Lemon peel effectively purified aqueous solutions by removal of Zn (II), Pb (II), Ni (II), Cd (II) and Cu (II) ions (Thirumavalavan et al. 2010). In other studies, the same scientific team used an orange peel. It showed effectiveness in removal of Cd (II), Cu (II), Ni (II), Pb (II) and Zn (II) ions (Thirumavalavan et al. 2010).

The presence of the citrus peel from orange was also studied by Ajmal et al. (2000). They obtained significant results for Ni (II) removal. Abdelhafez and Li (2016) removed significant amounts of Pb (II) with orange peel. Metal ions Pb (II), Ni (II), Zn (II), Cu (II) and Co (II) were effectively adsorbed by orange peel (Annadurai et al. 2002). On the other hand, removal of the metal ions Cd (II), Cu (II), Pb (II) and Ni (II) effectively increased after appropriate chemical modification of the orange skin (Feng et al. 2009; 2011).

Another agricultural waste, banana peels, has a significantly increased adsorptive capacity for Zn (II), Pb (II), Cd (II), Cu (II) and Ni (II) metals (Thirumavalavan et al. 2010). Subjecting banana peels to a pyrolysis process increased adsorptive possibilities for Cu (II) (DeMessie et al. 2015). Similar research using banana skin was conducted by Annadurai et al. (2002). He did not use any modifications in the process of removing heavy metals and received negligible adsorption capacity for metal ions Pb (II), Ni (II), Zn (II), Cu (II) and Co (II).

Cd (II) and Ni (II) were removed from the fluid solution by using a grapefruit peel. The bioremediation capacity of grapefruit skins is due to the ongoing ion exchange process and to the mechanism of complexation of metal ions with hydroxyl groups -OH (Torab-Mostaedi et al. 2013).

Villaescusa et al. (2004) used waste from grape stems in their studies. They found adsorptive capacity to eliminate Cu (II) and Ni (II).

Other products from agricultural production were also tested. Adsorption properties corn cob were found to include removal of heavy metal ions, specifically Cd (II). Chemical modification of maize cobs, caused by the addition of nitric acid and citric acid, increased the adsorptive potential tenfold (Leyva-Ramos et al. 2005). Corn cob was also used successfully to remove Pb (II) ions. The addition of sodium hydroxide (NaOH) caused a threefold increase in the adsorptive capacity (Tan et al. 2010).

Numerous studies on removing heavy metals from habitats were carried out using fungi. Li et al. (2018) used residues of four fungal species in studies: *Auricularia polytricha, Pleurotus eryngii, Flammulina velutipes and Pleurotus ostreatus*. These fungi species effectively accumulated Zn (II), Hg (II) and Cu (II) especially *F. velutipes*. However, Pb (II) and Cd (II) were most effectively removed by three species of fungi: *Agaricus bisporus* (button mushroom), *Pleurotus platypus* (oyster mushroom) and *Calocybe indica* (milky mushroom) (Vimala i Das 2009).

Most hat mushrooms have the ability to accumulate various heavy metals in the fruiting body. Heavy metal absorption processes from substrate to internal transport and accumulation in the fruiting body are dependent on many factors including genetic and environmental (climate, soil, and anthropogenic factors).

SOIL AND MINERAL DEPOSITS

Different degrees of affinity characterize soil and mineral deposits for removing metals from natural habitats. The affinity of soil for heavy metal is determined by the solubility of a heavy metal, charge, density, electronegativity or constant hydrolysis (Appel et al. 2008). Mineral deposits and soil properties make soil an ideal and effective material for removing heavy metals. Tests were carried out in which three soil samples were taken, and their capacity to remove heavy metals was evaluated. It turned out that soil effectively absorbed Pb (II) and Cd (II) (Appel et al. 2008). Kul and Koyuncu (2010) in their research used native and activated bentonite to remove Pb (II). The result showed that natural bentonite has a tenfold higher adsorptive capacity than activated bentonite (Kul and Koyuncu 2010).

The properties of peat moss depends mainly on the species composition of plant communities, the degree of humidification and peat bog irrigation conditions resulting from hydrogeological, hydrographic and topographic conditions. Qin et al. (2006), based on knowledge of peat structure indicated the presence of the carboxyl and phenolic functional groups As being conductive to heavy metal removal. They concluded that the existence of these functional groups increases the assimilation capacity of peat for removal of Cd (II), Cu (II) and Pb (II) ions from the environment. Pb (II) was removed with the highest efficiency (Ringqvist and Oborn 2002; Qin et al. 2006).

From other soils, natural kaolin was also used for research. The interaction between Pb (II) ions and the natural carbonate kaolin was used. In addition, the presence of a negative charge on the surface of the kaolin structure had an impact on the high adsorptive capacity of heavy metals (Tang et al. 2009).

Mineral deposits were also used in research to determine their adsorption capacity for heavy metal ions. Tiede et al. (2007) in their research used a mineral that contained manganese-oxyhydroxides. They used these minerals as a filter material to remove heavy metals from drinking water. The most effective removal was found for the ions Cd (II), then Zn (II) and trace amounts of Ni (II) ions.

In other studies, phosphate rock from ore was used. It has been shown to be effective in the removal of Cu (II), Pb (II), Cd (II) and Zn (II). In the next step, the raw material was subjected to chemical treatment with sodium hydroxide and nitric acid. This process increased the adsorption capacity by half (Elouear et al. 2008).

Wang and Mulligan (2006) compiled many available natural processes that are used to remove arsenic. Also, Kumpiene et al. (2008), reviewed methods of using different types of soil used for remediation. These methods are mainly intended to inactivate harmful ions and thereby neutralize their impact on the natural habitat. Jimenez-Castaneda and

Medina (2017) presented a study in which they used zeolites and clays to adsorb harmful ions from fluid. Using the influence of surfactants, they increased the adsorption capacity.

Finally, Derakhshan Nejad et al. in 2018 presented a comprehensive review of the use of various types of soils and mineral deposits. They show the abilities of mud and phosphate rocks to remove heavy metals from water sources.

There is much interest in using locally available soils and mineral deposits for remediation. These methods are primarily generally available and cheap to use.

OTHER WASTE MATERIAL

Industrial waste from the treatment of other materials has also been tested for effectiveness in eliminating harmful ions from natural habitats.

Raw material from black tea production, tea waste, has been repeatedly tested for its high removal efficiency of various heavy metals. Malkoc and Nuhoglu (2005) studied the influence of tea waste on Ni (II) ions, while Wasewar et al. (2009) found it to be effective in removing Zn (II) ions. Chemical modifications, such as acid and base rinsing, water vapor and ultrasound, significantly increased tea waste adsorption capacity for Cu (II) removal (Weng et al., 2014). Weng et al. 2014 put forward the hypothesis that high adsorption capacity is caused by the chemical modifications used. The surface area and porosity of the raw material was artificially increased which led to an increase in the number of functional groups.

Wastes from the production of green tea also showed the ability to remove heavy metals, in particular Ni (II) and As (III). It was observed that chemical modifications in the form of immersion in calcium hydroxide clearly increased adsorption possibilities (Yang et al., 2016). Tea waste was the subject of the work of Hussain et al. (2018), in which researchers presented a broad context for the suitability of waste in order to reduce heavy metals.

The waste material from coffee production was also tested. The coffee residues used showed adsorptive ability to remove various heavy metals, including Cu (II), Cd (II), Pb (II), Ni (II) and Zn (II) (Boonamnuayvitaya et al., 2004).

In other studies, Meunier et al. (2003) evaluated the adsorption capacity of cocoa cover dissolved in solution. Cocoa shell removed of a wide spectrum of heavy metals includingCd (II), Co (II), Cr (III), Cu (II), Fe (III), Mn (II), Ni (II), Pb (II) and Zn (II) (Meunier et al. 2003).

Qi and Aldrich (2008) tested tobacco dust dissolved in an aqueous solution. It turned out that tobacco dust removes Cu (II), Zn (II), Cd (II), Pb (II) and Ni (II) ions. Researchers explained that tobacco acidified solution, which together with negatively charged metal ions, had a positive effect on the eliminating of harmful ions (Qi and Aldrich, 2008).

The raw material from the production of olive oil was shown to be effective in removal of Cd (II) (Al-Anber and Matouq, 2008).

Ashes from the combustion of coal can be used for its adsorptive capacity to eliminate heavy metals from industrial facility or municipal sludge waste.An important factor for the effectiveness of eliminating heavy metals from liquid fluid using ashes is its ability to alkalize solutions. Removal of heavy metals using fly ash has a selective character, especially for low pH values of the solution (Kasprzyk and Dyjakon 2017). Papandreou et al. (2011), with the help of ash from coal combustion, removed Pb (II), Zn (II) and Cr (III) ions. This phenomenon is explained by the electrostatic attraction between metal ions and the adsorbent surface that has a positive charge (Papandreou et al. 2011). Also coal fly ash was used to remove Hg (II) ions from industrial wastewater. A high 95% removal efficiency was obtained (Attari et al. 2017).

Many review articles were published which discuss the high adsorption potential of coal ash as an adsorbent (Cho et al. 2005; Wang and Wu 2006; Varma et al. 2013; Rashidi and Yusup 2016; Ge et al. 2018).

Shells of various species of aquatic mollusks effectively absorb heavy metals. *Anadara inaequivalvis* shells fished out of the Adriatic, Aegean and Black Sea in fact eliminate Pb (II) and Cu (II) (Bozbas and Boz 2016). Research conducted on aquatic mollusks by Du et al. (2011) showed significant adsorption capabilities. Husband shells accumulated Pb (II), Cd (II) and Zn (II), while oyster shells removed Pb (II), Zn (II) and Cd (II) (Du et al. 2011).

Vijayaraghavan et al. (2006) demonstrated the bioremediation capacity of crab shell particles, which removed Co (II) and Cu (II) ions.

Dakiky et al. (2002) showed that wool had the highest adsorption ability for Cr (VI) ions compared to olives, cactus leaves and carbon. The high efficiency of wool was explained by its loose structure in contrast to other adsorbents that are characterized by a compact structure (Dakiky et al. 2002).

CONCLUSION

Environmental pollution is a significant problem. There are various forms of pollution, but heavy metals are particularly dangerous for the environment and the health and life of animals including humans.

Bearing in mind the danger of heavy metals, many studies have been carried out using adsorbents to remove them from the environment. The most frequently tested raw materials are from readily available sources i.e., agricultural waste, soil, mineral deposits, water and soil biomass.

Numerous studies confirm that raw materials such as agricultural waste and industrial by-products are very effective when used as adsorbents. Chemical modifications of the tested raw materials usually increase their adsorption capacity.

Many scientific articles emphasize methods and technologies that can be used to eliminate heavy metals. These are effective methods and, most importantly, can be used for many developed and developing countries. Membrane filtration (Kim et al. 2018), electrocoagulation (Al-Qodah and Al-Shannag 2017; Bazrafshan et al. 2015), microbiological remediation (Ayangbenro and Babalola 2017; Li and Tao 2015), adsorption of activated carbon (Li et al. 2018; Renu et al. 2017), carbon nanotechnology (Peng et al. 2017; Sherlala et al. 2018; Xu et al. 2018) and various chemical modifications of adsorbents (Jiang et al. 2018; Sajida et al. 2018; Zare et al. 2018) are the most effective methods.

Research in this topic should be continued to identify new available raw materials and to popularize and simplify methods that will be useful in the eliminate of harmful ions.

ACKNOWLEDGMENTS

Dorota Bartusik-Aebisher acknowledges support from the National Center of Science NCN (New drug delivery systems-MRI study, Grant OPUS-13 number 2017/25/B/ST4/02481).

REFERENCES

Abdelhafez AA, Li J (2016) Removal of Pb(II) from aqueous solution by using biochars derived from sugar cane bagasse and orange peel. *J Taiwan Inst Chem Eng* 61:367-375.

Ahmad A, Ghazi ZA, Saeed M, Ilyas M, Ahmad R, Khattaka AM, Iqbal A (2017) A comparative study of the removal of Cr(VI) from synthetic solution using natural biosorbents. *New J Chem* 41:10799-10807.

Ajjabi LC, Chouba L (2009) Biosorption of Cu(2+) and Zn(2+) from aqueous solutions by dried marine green macroalga *Chaetomorpha linum*. *J Environ Manage* 90(11):3485-3489.

Ajmal M, Rao RA, Ahmad R, Ahmad J (2000) Adsorption studies on citrus reticulata (fruit peel of orange): removal and recovery of Ni(II) from electroplating wastewater. *J Hazard Mater* 79(1-2):117-131.

Ajmal M, Rao RA, Anwar S, Ahmad J, Ahmad R (2003) Adsorption studies on rice husk: removal and recovery of Cd(II) from wastewater. *Bioresour Technol* 86(2):147-149.

Al-Anber ZA, Matouq, MA (2008) Batch adsorption of cadmium ions from aqueous solution by means of olive cake. *J Hazard Mater* 151(1):194-201.

Al-Qodah Z, Al-Shannag M (2017) Heavy metal ions removal from wastewater using electrocoagulation processes: a comprehensive review. *Separ Sci Technol* 52(17):2649-2676.

Annadurai G, Juang RS, Lee DJ (2002) Adsorption of heavy metals from water using banana and orange peels. *Water Sci Technol* 47(1): 185-190.

Apiratikul R and Pavasant P (2006) Sorption isotherm model for binary component sorption of copper, cadmium, and lead ions using dried green macroalga, *Caulerpa lentillifera*. *Chem Eng J* 119(2–3):135-145.

Appel C, Ma LQ, Rhue RD, Reve W (2008) Sequential sorption of lead and cadmium in three tropical soils. *Environ Pollut* 155(1):132-140.

Archundia D, Duwig CC, Spadini L, Uzu G, Guédron S, Morel MC, Cortez R, Ramos O, Chincheros J, Martins JMF (2017). How uncontrolled urban expansion increases the contamination of the titicaca lake basin (El Alto, La Paz, Bolivia). *Water Air Soil Pollut* 228:44-60.

Attari M, Bukhari S, Kazemian H, Rohani S (2017) A low-cost adsorbent from coal fly ash for mercury removal from industrial wastewater. *J Environ Chem Eng* 5:391-399.

Ayangbenro AS, Babalola OO (2017) A new strategy for heavy metal polluted environments: a review of microbial biosorbents. *Int J Environ Res Public Health* 14(1):94-109.

Bansal M, Garg U, Singh D, Garg VK (2009) Removal of Cr(VI) from aqueous solutions using pre-consumer processing agricultural waste: a case study of rice husk. *J Hazard Mater* 162:312-320.

Bansode RR, Losso JN, Marshall WE, Rao RM, Portier RJ (2003) Adsorption of metal ions by pecan shell-based granular activated carbons. *Bioresour Technol* 89(2):115-119.

Bazrafshan E, Mohammadi L, Ansari-Moghaddam A, Mahvi AH (2015) Heavy metals removal from aqueous environments by electrocoagulation process – a systematic review. *J Environ Health Sci Eng* 13: 74-90.

Bhattacharya A, Tiwari P, Sahu PK, Kumar S (2018) Review of the Phytochemical and Pharmacological Characteristics of *Moringa oleifera*. *J Pharm Bioallied Sci.* 10(4):181–191.

Boonamnuayvitaya V, Chaiya C, Tanthapanichakoon W, Jarudilokkul S (2004) Removal of heavy metals by adsorbent prepared from pyrolyzed coffee residues and clay. *Separ Purif Technol* 35:11-22.

Bozbas SK, Boz Y (2016) Low-cost biosorbent: Anadara inaequivalvis shells for removal of Pb(II) and Cu(II) from aqueous solution. *Process Saf Environ Protect* 103:144-152.

Bozic D, Stankovic V, Gorgievski M, Bogdanovic G, Kovacevic R (2009) Adsorption of heavy metal ions by sawdust of deciduous trees. *J Hazard Mater* 171:684-692.

Chakraborty S, Dutta AR, Sural S, Gupta D, Sen S (2013) Ailing bones and failing kidneys: a case of chronic cadmium toxicity. *Ann Clin Biochem* 50:492-495.

Chen H, Zhao J, Dai G, Wu J, Yan H (2010) Adsorption characteristics of Pb(II) from aqueous solution onto a natural biosorbent, fallen *Cinnamomum camphora* leaves. *Desalination* 262:174-182.

Chen Z, Ma W, Han M (2008) Biosorption of nickel and copper onto treated alga (*Undaria pinnatifida*): application of isotherm and kinetic models. *J Hazard Mater* 155(1-2):327-333.

Cho H, Oh D, Kim K (2005) A study on removal characteristics of heavy metals from aqueous solution by fly ash. *J Hazard Mater* 127(1-3):187-195.

Dakiky M, Khamis M, Manassra A, Mer'eb M (2002) Selective adsorption of chromium(VI) in industrial wastewater using low-cost abundantly available adsorbents. *Adv Environ Res* 6:533-540.

DeMessie B, Sahle-Demessie E, Sorial G (2015) Cleaning water contaminated with heavy metal ions using pyrolyzed biochar adsorbents. *Separ Sci Technol* 50(16):2448-2457.

Demirak A, Yilmaz F, Tuna AL, Ozdemir N (2006) Heavy metals in water, sediment and tissues of *Leuciscus cephalus* from a stream in southwestern Turkey. *Chemosphere* 63(9):1451-1458.

Demirbas E, Dizge N, Sulak MT, Kobya M (2009) Adsorption kinetics and equilibrium of copper from aqueous solutions using hazelnut shell activated carbon. *Chem Eng J* 148:480-487.

Deng L, Su Y, Su H, Wang X, Zhu X (2006) Biosorption of copper (II) and lead (II) from aqueous solutions by nonliving green algae *Cladophora fascicularis*: Equilibrium, kinetics and environmental effects. *Adsorption* 12(4):267–277.

Derakhshan Nejad Z, Jung MC, Kim KH (2018) Remediation of soils contaminated with heavy metals with an emphasis on immobilization technology. *Environ Geochem Health* 40(3):927-953.

Du Y, Lian F, Zhu L (2011) Biosorption of divalent Pb, Cd and Zn on aragonite and calcite mollusk shells. *Environ Pollut* 159(7):1763-1768.

Elangovan R, Philip L, Chandraraj K (2008). Biosorption of chromium species by aquatic weeds: kinetics and mechanism studies. *J Hazard Mater* 152(1):100-112.

Elouear Z, Bouzid J, Boujelben N, Feki M, Jamoussi F, Montiel A (2008) Heavy metal removal from aqueous solutions by activated phosphate rock. *J Hazard Mater* 156(1-3):412-420.

El-Sikaily A, El Nemr A, Khaled A, Abdelwehab O (2007) Removal of toxic chromium from wastewater using green alga *Ulva lactuca* and its activated carbon. *J Hazard Mater* 148(1-2):216-228.

Feng N, Guo X, Liang S (2009) Adsorption study of copper(II) by chemically modified orange peel. *J Hazard Mater* 164(2-3):1286-1292.

Feng N, Guo X, Liang S, Zhu Y, Liu J (2011) Biosorption of heavy metals from aqueous solutions by chemically modified orange peel. *J Hazard Mater* 185(1):49-54.

Ge J, Yoon S, Choi N (2018) Application of fly ash as an adsorbent for removal of air and water pollutants. *Appl Sci* 8:1116-1139.

Ge Y, Li Z (2018) Application of lignin and its derivatives in adsorption of heavy metal ions in water: A review. ACS Sustain. *Chem Eng* 6: 7181-7192.

Georgieva VG, Tavlieva MP, Genieva SD, Vlaev LT (2015) Adsorption kinetics of Cr(VI) ions from aqueous solutions onto black rice husk ash. *J Mol Liq* 208:219-226.

Gleason KM, Valeri L, Shankar AH, Hasan M, Quamruzzaman Q, Rodrigues EG, Christiani DC, Wright RO, Bellinger DC, Mazumdar M (2016) Stunting is associated with blood lead concentration among Bangladeshi children aged 2-3 years. *Environ Health* 15:103-111.

Guo X, Zhang S, Shan XQ (2008) Adsorption of metal ions on lignin. *J Hazard Mater* 151(1):134-142.

Hashim M, Chu K (2004) Biosorption of cadmium by brown, green, and red seaweeds. *Chem Eng J* 97:249-255.

Herat S, Agamuthu P (2012) E-waste: a problem or an opportunity? Review of issues, challenges and solutions in Asian countries. *Waste Manag Res* 30(11):1113-1129.

Ho YS, Huang CT, Huang HW (2002) Equilibrium sorption isotherm for metal ions on tree fern. *Process Biochem* 37:1421-1430.

Hussain S, Anjali K, Hassan S, Dwivedi P (2018) Waste tea as a novel adsorbent: a review. *Appl Water Sci* 8:165-180.

Jaishankar M, Tseten T, Anbalagan N, Mathew BB, Beeregowda KN (2014) Toxicity, mechanism and health effects of some heavy metals. *Interdiscip Toxicol.* 7(2):60-72.

Jarup L (2003). Hazards of heavy metal contamination. *Br Med Bull* 68:167e182.

Jiang Y, Liu Z, Zeng G, Liu Y, Shao B, Li Z, Liu Y, Zhang W, He Q (2018) Polyaniline-based adsorbents for removal of hexavalent chromium from aqueous solution: a mini review. *Environ Sci Pollut Res* 25:6158-6174.

Jimenez-Castaneda M, Medina D (2017) Use of surfactant-modified zeolites and clays for the removal of heavy metals from water. *Water* 9:235-246.

Kalibbala HM, Wahlberg O, Hawumba TJ (2009) The impact of *Moringa oleifera* as a coagulant aid on the removal of trihalomethane (THM) precursors and iron from drinking water". *Wat Sci Tech. Water Supply* 9(6):707–714.

Kansal SK, Kumari A (2014) Potential of *M. oleifera* for the treatment of water and wastewater. *Chem Rev* 114(9):4993-5010.

Kasprzyk K, Dyjakon A (2017) The use of fly ashes for heavy metals removal from sewage sludges assigned to agricultural field fertilization. *Zesz Probl Post Nauk Roln* 589:27–37.

Kim S, Chu K, Al-Hamadani Y, Park C, Jang M, Kim D, Yu M, Heo J, Yoon Y (2018) Removal of contaminants of emerging concern by membranes in water and wastewater: a review. *Chem Eng J* 335:896-914.

Krishnani KK, Meng X, Christodoulatos C, Boddu VM (2008) Biosorption mechanism of nine different heavy metals onto biomatrix from rice husk. *J Hazard Mater* 153(3):1222-1234.

Kul AR, Koyuncu H (2010) Adsorption of Pb(II) ions from aqueous solution by native and activated bentonite: kinetic, equilibrium and thermodynamic study. *J Hazard Mater* 179(1-3)-332-339.

Kumpiene J, Lagerkvist A, Maurice C (2008) Stabilization of As, Cr, Cu, Pb and Zn in soil using amendments - a review. *Waste Manag* 28:215-225.

Leyva-Ramos R, Bernal-Jacome LA, Acosta-Rodriguez I (2005) Adsorption of cadmium(II) from aqueous solution on natural and oxidized corncob. *Separ Purif Technol* 45:41-49.

Li J, Zheng B, He Y, Zhou Y, Chen X, Ruan S, Yang Y, Dai C, Tang L (2018) Antimony contamination, consequences and removal techniques: a review. *Ecotoxicol Environ Saf* 156:125-134.

Li PS, Tao HC (2015) Cell surface engineering of microorganisms towards adsorption of heavy metals. *Crit Rev Microbiol* 41(2):140-149.

Li Q, Zhai J, Zhang W, Wang M, Zhou J (2007) Kinetic studies of adsorption of Pb(II), Cr(III) and Cu(II) from aqueous solution by sawdust and modified peanut husk. *J Hazard Mater* 141(1):163-167.

Li X, Zhang D, Sheng F, Qing H (2018) Adsorption characteristics of copper(II), zinc(II) and mercury(II) by four kinds of immobilized fungi residues. *Ecotoxicol Environ Saf* 147:357-366.

Liang F, Song Y, Huang C, Zhang J, Chen B (2013) Adsorption of hexavalent chromium on a lignin-based resin - equilibrium, thermodynamics, and kinetics. *J Environ Chem Eng* 1:1301-1308.

Malkoc E, Nuhoglu Y (2005) Investigations of nickel(II) removal from aqueous solutions using tea factory waste. *J Hazard Mater* B 127: 120-128.

Masindi V; Muedi KL (2018) Environmental Contamination by Heavy Metals. In *Heavy Metals* edited by Hosam El-Din M. Saleh, published: June 27th 2018; DOI: 10.5772/intechopen.71185.

Meitei MD, Prasad M (2014) Adsorption of Cu(II), Mn(II) and Zn(II) by *Spirodela polyrhiza* (L.) Schleiden: equilibrium, kinetic and thermodynamic studies. *Ecol Eng* 71:308-317.

Neris JB, Luzardo FHM, da Silva EGP, Velasco FG (2019) Evaluation of adsorption processes of metal ions in multi-element aqueous systems by lignocellulosic adsorbents applying different isotherms: a critical review. *Chem Eng J* 357:404-420.

Meunier N, Laroulandie J, Blais JF, Tyagi RD (2003) Cocoa shells for heavy metal removal from acidic solutions. *Bioresour Technol* 90(3):255-263.

Moussavi G, Barikbin B (2010) Biosorption of chromium(VI) from industrial wastewater onto pistachio hull waste biomass. *Chem Eng J* 162:893-900.

Olafisoye OB, Adefioye T, Osibote OA (2013) Heavy metals contamination of water, soil, and plants around an electronic waste dumpsite. *Pol J Environ Stud* 22(5):1431-1439.

Onyancha D, Mavura W, Ngila JC, Ongoma P, Chacha J (2008) Studies of chromium removal from tannery wastewaters by algae biosorbents, *Spirogyra condensata* and *Rhizoclonium hieroglyphicum*. *J Hazard Mater* 158(2-3):605-614.

Pandey J, Pandey U (2009) Atmospheric deposition and heavy metal contamination in an organic farming system in a seasonally dry tropical region of India. *J Sustain Agric* 33:361-378.

Papandreou AD, Stournaras CJ, Panias D, Paspaliaris I (2011) Adsorption of Pb(II), Zn(II) and Cr(III) on coal fly ash porous pellets. *Miner Eng* 24:1495-1501.

Peng W, Li H, Liu Y, Song S (2017) A review on heavy metal ions adsorption from water by graphene oxide and its composites. *J Mol Liq* 230:496-504.

Perkins DN, Brune Drisse MN, Nxele T, Sly PD (2014) E-waste: a global hazard. *Ann Glob Health* 80(4):286-295.

Prasse C, Zech W, Itanna F, Glaser B (2012) Contamination and source assessment of metals, polychlorinated biphenyls, and polycyclic aromatic hydrocarbons in urban soils from Addis Ababa, Ethiopia. *Toxicol Environ Chem* 94(10):1954-1979.

Qi BC, Aldrich C (2008) Biosorption of heavy metals from aqueous solutions with tobacco dust. *Bioresour Technol* 99(13):5595-5601.

Qin F, Wen B, Shan XQ, Xie YN, Liu T, Zhang SZ, Khan SU (2006) Mechanisms of competitive adsorption of Pb, Cu, and Cd on peat. *Environ Pollut* 144(2):669-680.

Rashidi N, Yusup S (2016) Overview on the potential of coal-based bottom ash as low-cost adsorbents. *ACS Sustain Chem Eng* 4(4):1870-1884.

Renu Agarwal M, Singh K (2017) Heavy metal removal from wastewater using various adsorbents: a review. *J Water Reuse Desal* 7(4):387-419.

Ringqvist L, Oborn I (2002) Copper and zinc adsorption onto poorly humified *Sphagnum* and *Carex* peat. *Water Res* 36(9):2233-2242.

Romera E, González F, Ballester A, Blázquez M, Muñoz J (2008) Biosorption of Cd, Ni, and Zn with mixtures of different types of algae. *Environ Eng Sci* 25(7):999-1008.

Sajid M, Nazal MK, Ihsanullah, Baig N, Osman AM (2018) Removal of heavy metals and organic pollutants from water using dendritic polymers based adsorbents: a critical review. *Separ Purif Technol* 191:400-423.

Schwartzbord JR, Emmanuel E, Brown DL (2013) Haiti's food and drinking water: a review of toxicological health risks. *Clin Toxicol.* 51 (9):828-833.

Senthil Kumar PS, Ramalingam S, Dinesh Kirupha SD, Murugesan A, Vidhyadevi T, Sivanesan S (2011) Adsorption behavior of nickel(II) onto cashew nut shell: equilibrium, thermodynamics, kinetics, mechanism and proces design. *Chem Eng J* 167:122-131.

SenthilKumar P, Ramalingam S, Sathyaselvabala V, Dinesh Kirupha S, Sivanesan S (2011) Removal of copper(II) ions from aqueous solution by adsorption using cashew nut shell. *Desalination* 266:63-71.

Shan T, Matar M, Makky E, Ali E (2017) The use of *Moringa oleifera* seed as a natural coagulant for wastewater treatment and heavy metals removal. *Appl Water Sci* 7, 1369-1376.

Sherlala A, Raman A, Bello M, Asghar A (2018) A review of the applications of organo-functionalized magnetic graphene oxide nanocomposites for heavy metal adsorption. *Chemosphere* 193: 1004-1017.

Shukla SR, Pai RS (2005) Adsorption of Cu(II), Ni(II) and Zn(II) on dye loaded groundnut shells and sawdust. *Separ Purif Technol* 43(1):1-8.

Singha B, Das SK (2013) Adsorptive removal of Cu(II) from aqueous solution and industrial effluent using natural/agricultural wastes. *Colloids Surfaces B Biointerfaces* 107:97-106.

Srivastava S, Agrawal SB, Mondal MK (2015) Biosorption isotherms and kinetics on removal of Cr(VI) using native and chemically modified *Lagerstroemia speciosa* bark. *Ecol Eng* 85:56-66.

Tan G, Yuan H, Liu Y, Xiao D (2010) Removal of lead from aqueous solution with native and chemically modified corncobs. *J Hazard Mater* 174:740-745.

Tang Q, Tang X, Li Z, Chen Y, Kou N, Sun Z (2009) Adsorption and desorption behaviour of Pb(II) on a natural kaolin: equilibrium, kinetic, and thermodynamic studies. *J Chem Technol Biotechnol* 84:1371-1380.

Taşar S, Kaya F, Özer, A (2014) Biosorption of lead(II) ions from aqueous solution by peanut shells: equilibrium, thermodynamic and kinetic studies. *J Environ Chem Eng* 2(2):1018:1026.

Tchounwou PB, Yedjou CG, Patlolla AK, Sutton DJ (2012) Heavy Metals Toxicity and the Environment. *EXS* 101:133–164.

Thirumavalavan M, Lai Y, Lin L, Lee J (2010) Cellulose-based native and surface modified fruit peels for the adsorption of heavy metal ions from aqueous solution. *J Chem Eng Data* 55:1186-1192.

Tiede K, Neumann T, Stueben D (2007) Suitability of Mn-oxyhydroxides from karst caves as filter material for drinking water treatment in Gunung Sewu, Indonesia. *J Soils Sediments* 7(1):53-58.

Torab-Mostaedi M, Asadollahzadeh M, Hemmati A, Khosravi (2013) Equilibrium, kinetic, and thermodynamic studies for biosorption of cadmium and nickel on grapefruit peel. *J Taiwan Inst Chem Eng* 44:295-302.

Tóth G, Hermann T, Da Silva MR, Montanarella L (2016) Heavy metals in agricultural soils of the European Union with implications for food safety. *Environ Int* 88:299-309.

Varma GV, Karan Singh R, Sahu V (2013) A comparative study on the removal of heavy metals by adsorption using fly ash and sludge: A review. *IJAIEM* 2(7):45-56.

Vijayaraghavan K, Palanivelu K, Velan M (2006) Biosorption of copper(II) and cobalt(II) from aqueous solutions by crab shell particles. *Bioresour Technol* 97(12):1411-1419.

Villaescusa I, Fiol N, Martinez M, Miralles N, Poch J, Serarols J (2004) Removal of copper and nickel ions from aqueous solutions by grape stalks wastes. *Water Res* 38(4):992-1002.

Vimala R, Das N (2009) Biosorption of cadmium(II) and lead(II) from aqueous solutions using mushrooms: a comparative study. *J Hazard Mater* 168:376-382.

Wang S, Mulligan CN (2006) Natural attenuation processes for remediation of arsenic contaminated soils and groundwater. *J Hazard Mater* 138(3):459-470.

Wang S, Wu H (2006) Environmental-benign utilisation of fly ash as low-cost adsorbents. *J Hazard Mater* 136(3):482-501.

Wasewar KL, Atif M, Prasad B, Mishra IM (2009) Batch adsorption of zinc on tea factory waste. *Desalination* 244:66-71.

Weng C, Lin Y, Hong D, Sharma YC, Chen S, Tripathi K (2014) Effective removal of copper ions from aqueous solution using base treated black tea waste. *Ecol Eng* 67:127-133.

Witek-Krowiak A, Szafran RG, Modelski S (2011) Biosorption of heavy metals from aqueous solutions onto peanut shell as a low-cost biosorbent. *Desalination* 265:126-134.

Wu Q, Leung JY., Geng X, Chen S, Huang X, Li H, Huang Z, Zhu L, Chen J, Lu Y (2015) Heavy metal contamination of soil and water in the vicinity of an abandoned e-waste recycling site: implications for dissemination of heavy metals. *Sci Total Environ* 506-507:217-225.

Wu Y, Zhang S, Guo X, Huang H (2008) Adsorption of chromium(III) on lignin. *Bioresour Technol* 99(16):7709-7715.

Xu J, Cao Z, Zhang Y, Yuan Z, Lou Z, Xu X, Wang X (2018) A review of functionalized carbon nanotubes and graphene for heavy metal adsorption from water: Preparation, application, and mechanism. *Chemosphere* 195:351-364.

Yang S, Wu Y, Aierken A, Zhang M, Fang P, Fan Y, Ming Z (2016) Mono/competitive adsorption of arsenic(III) and nickel(II) using modified green tea waste. *J Taiwan Inst Chem Eng* 60:213-221.

Yu LJ, Shukla SS, Dorris KL, Shukla A, Margrave JL (2003) Adsorption of chromium from aqueous solutions by maple sawdust. *J Hazard Mater* 100:53-63.

Zare EN, Motahari A, Sillanpää M (2018) Nanoadsorbents based on conducting polymer nanocomposites with main focus on polyaniline and its derivatives for removal of heavy metal ions/dyes: a review. *Environ Res* 162:173-195.

Zhang M (2011) Adsorption study of Pb(II), Cu(II) and Zn(II) from simulated acid mine drainage using dairy manure compost. *Chem Eng J* 172:361-368.

Zhu CS, Wang LP, Chen WB (2009) Removal of Cu(II) from aqueous solution by agricultural by-product: peanut hull. *J Hazard Mater* 168(2-3):739-746.

In: Environmental Science of Heavy Metals ISBN: 978-1-53617-831-9
Editor: Dorota Bartusik-Aebisher © 2020 Nova Science Publishers, Inc.

Chapter 4

BIOLOGICAL STRATEGIES FOR HEAVY METAL REMOVAL

Sabina Galiniak, Tomasz Kubrak, Rafał Podgórski,
David Aebisher and Dorota Bartusik-Aebisher[*]
Faculty of Medicine, University of Rzeszow, Poland

ABSTRACT

Recently, there has been a growing interest in technology called phytoremediation which includes the employment of living plant organisms for cleaning the environment. Phytoremediation technologies rely on the cultivation of species of plants that are able to grow in a contaminated environment. The manner in which the plant interacts with heavy metals depends mainly on the type of contamination, species of plant as well as conditions. Few plant species have the capability to accumulate metals in their own cells, others are able to incorporate them into their metabolic pathways. Moreover, thanks to special chemical compounds secreted by roots, these plants can cause binding of harmful substances which leads to limiting their flow into soil. These processes

---

[*] Corresponding Author's Email: dbartusik-aebisher@ur.edu.pl.

are the result of the natural adaptation of plants to the prevailing conditions, and these plants are treated as bioindicators of selected elements.

The chapter presents biological strategies for the elimination of heavy metals from polluted habitats including processes of phytoextraction, phytostabilization, phytodegradation, phytostimulation, phytovolatilization and phytofiltration.

Keywords: hyperaccumulators, phytoremediation, phytoextraction, phytostabilization, phytodegradation, phytostimulation, phytovolatilization, phytofiltration

INTRODUCTION

Anthropogenic activities have an adverse influence on the ecosystem that leads to pollution and degradation. The main sources of contamination, with an extensive spectrum of compositions and concentrations, are industry and agriculture, which mainly release sewage with inorganic and organic compounds containing heavy metals into the environment. Soil may become contaminated by emission from industrial areas, mine tailings, automobile exhaust, urban runoff, animal manures, plant protection agents, and fertilizers (Wuana and Okieimen 2011; Yang et al. 2018).

Among the most commonly found heavy metals in contaminated grounds are arsenic (As), cadmium (Cd), chromium (Cr), copper (Cu), lead (Pb), mercury (Hg), nickel (Ni), and zinc (Zn). According to report by Chen et al. (2016), in the past 50 years, at least 30 thousand tons of Cr and 800 tons of Pb were discharged into the ecosystem, most of which into soil.

Currently, due to severe soil pollution and the harmful impacts of heavy metals, there are high expectations in the use of biological methods as efficient processes for purification of the environment. Among these methods, phytoremediation (also known as green remediation, botanoremediation or vegetative remediation) are beginning to play an increasingly important role where plant activity involving the uptake and degradation of environmental pollutants are employed.

Traditional physical and chemical procedures of heavy metal removal from a contaminated ecosystem are generally not adaptable to large areas and are usually expensive and not socially preferred. In contrast to these methods, phytoremediation is highly useful in extensively polluted sites, especially where the elimination of heavy metal contamination is not immediate.

According to a study by Salt et al. (1995), to remove pollutants from one acre of ground to a depth of 0.5 m, phytoremediation costs less than 1 million dollars, while physical remediation costs at least 4 times as much money.

The word phytoremediation originates from the Ancient Greek term *phyton* – plant and Latin – *remediare* – returning balance. Currently, phytoremediation is considered to be an effective, non-invasive, low-cost, socially acceptable and ecological technology when compare to other methods of remediation.

Due to the way in which plants affect the purification of contaminated ecosystems, main types of phytoremediation are distinguished on the basis of fundamental methods and applications as: phytoextraction, phytostabilization, phytodegradation, phytostimulation, phytovolatilization and phytofiltration. In practice, however, the mechanisms for removing or detoxifying particular impurities are more complex and often result from the combination of the various methods listed.

Toxic Impacts of Heavy Metals on Health

The toxicity of released heavy metals to water and soil is a complication with growing importance for not only environmental, but also for ecological and nutritional reasons. Moreover, exposure to excessive concentrations of heavy metals may contribute to human health problems. Heavy metals get in the human organism though food, water and air and when they are not metabolized, they accumulate as toxins in the tissue.

Studies on animals show that heavy metals have a various distribution in tissue, however, Cu, Pb and Hg mainly accumulate in the kidney and liver, and Zn in muscle tissue (Alkmim Filho et al. 2014; Ali et al. 2019; Andjelkovic et al. 2019). Moreover, accumulation of heavy metals in animal tissue is dependent on species as well as age (Rudy et al. 2007; Heba et al. 2015). A report by Li et al. (2015) presents the concentration of heavy metals in different rat tissue after exposure to smog particles – PM2.5 which contain high level of toxic metals, including Cr, Cd, Ni, Pb, As, and Zn. Histological analysis of tissues revealed that the maximum concentration of Pb was noted in liver, lung as well as the cerebral cortex. Likewise, a report by Pamphlett et al. (2018) showed that the concentration of Hg in the locus ceruleus increases with age and may be implicated in neurological disorders.

After absorption, heavy metals are spread in the body via the circulatory system, where they might be bound to high-molecular-weight proteins or hemoglobin (Timchalk et al. 2006). After reaching their destination, they enter cells mimicking natural ions and utilizing their transporters.

Exposure to various environmental pollutants is known to induce changes in various tissue and systems. Probable mechanisms of heavy metal harmfulness include production of reactive oxygen species (ROS) via Fenton or Fenton-type reactions, interaction with other elements, and impairment of enzyme activities, and cellular processes such as apoptosis and DNA repair (Matović et al. 2015; Zhou et al. 2017; Andjelkovic et al. 2019). Moreover, heavy metals cause severe consequences after interaction with DNA such as induction of conformational changes of the double helix, impairment of interaction of proteins with DNA, modification of nucleosome structure, changes in the degree of chromatin compaction, and induction of abnormalities in chromatin condensation (Sas-Nowosielska and Pawlas 2015). Persistent exposure of bronchial epithelial cells to As and Cr lead to malignant cell transformation and carcinogenesis. Analysis of cell changes revealed that treatment of cells with As caused reduction in the activity of p53 and p21, while in case of Cr, activity of these proteins were elevated (Park et al. 2015). Incubation of Cd with human peripheral blood mononuclear cells and B cells caused inhibition of cell proliferation and IgE synthesis (Marth et al. 2001). According to study by Keogh et al. (1994), rank order cytotoxicity in the human small intestinal epithelial cell line induced by heavy metals was found to be Hg > Cd > Cu > Pb. In general, heavy metals cause stomatitis, psychic tension, convulsions, lack of voluntary coordination of muscle movements, and loss of agility. Moreover, they lead to hemoglobinuria, pneumonia, neurotoxicity, nephrotoxicity, male infertility and even development of cancer and diabetes via insulin deficiency and death of pancreatic β cells (Vella et al. 2017; Rehman et al. 2018).

PLANTS USED FOR PHYTOREMEDIATION

General Characteristics

Plants suitable for phytoremediation processes are characterized by tolerance to elevated levels of toxic metals, a high degree of accumulation or biodegradation of impurities, even at a relatively low level of pollutions, the capability to adsorb several pollutions at the same time, fast growth and high biomass, resistance to diseases as well as difficult environmental conditions. However, plant species partially fulfill these conditions due to differentiated ability to incept and accumulate heavy metals. It seems that understanding the process of plants insusceptibility to a specific metal is crucial for their growth and use for phytoremediation at contaminated sites.

Storage and dissemination of heavy metals in plants hinges on various factors including the plant species, its tissues, metal type, its oxidation state, pH, concentration, vegetation period, features of the ground, and so forth (Al-Farraj and Al-Wabel 2007; Tangahu et al. 2011; Filipović-Trajković et al. 2012; Morkunas et al. 2018). The uptake of metals is mainly reached by releasing compounds related to metal chelation such as mugenic and aveic acids which cause binding of metals to soil. After entering the roots, heavy metals can be accumulated in the roots or bound to phytochelatin and transported to the shoot through xylem. Moreover, mobility of heavy metals is assisted by other molecules including malate, citrate, and histidine (Jabeen et al. 2009). Furthermore, enzymes called ATPases are the kay factors in the translocation or counteraction of toxic compounds. Studies on these transporters revealed that they are varied in terms of tissue distribution, subcellular localization, and metal specificity (Takahashi et al. 2012).

The capacity to resist increased concentration of heavy metals as well as storage them in elevated levels has been confirmed and described in many plants belonging to various species. The process of heavy metal absorption is steady during the vegetation period, however, the highest value of uptake is observed at the end of the vegetation period (Krstic et al. 2007). It is established that the content of these elements might be even several times larger in the tissue of wild vegetable organisms than in the cultivated ones (Filipović-Trajković et al. 2012). Analyses indicate that about 1% of the entire plant body stores these elements. A few of them are bound to protein and play essential roles in their structures and functions, especially in enzymatic proteins (Shallari et al. 1997). However, high concentration of these elements cause alteration in morphology, physiology and biochemistry in vegetable organisms (Farid et al. 2013; Shahid et al. 2014; Morkunas et al. 2018). Generally, it is established that exposure to heavy metals increase ROS generation which are responsible for impairment of molecules and, finally, cell components. Severe production of ROS also causes changes in the antioxidant system and oxidation of biomolecules. Furthermore, plant responses to heavy metals are noted as alternations in redox status, the level of molecules involved in signal transduction, membrane permeability, cysteine, glutathione and phytochelatin levels as well as expression of genes encoding enzymes of the reactions of flavonoid biosynthesis (Morkunas et al. 2018).

Exposure to heavy metals also leads to reduction in size of leaves and flowers, plant cell proliferation, water uptake and causes turgor loss via decreasing cell wall elasticity. Moreover, the toxicity of heavy metals leads to decline of photosynthetic rate that might cause many changes in the physiological abilities of plants and, consequently, plant growth and productivity (Farid et al. 2013; Morkunas et al. 2018).

Vegetable organisms require a balance between the absorption of necessary ions to keep regular plant physiology and the ability to protect easily harmed cellular processes and organelles from elevated levels of metals.

Moreover, protection of plants from excessive metal ions might be reached by prevention methods, which primarily relies on the binding of metal in root and components of plant wall.

Tolerance to heavy elements is based on the accumulation in vacuoles, and by attachment to specific compounds like organic acids and peptides and on the occurrence of enzymatic reactions adapted to increased levels of heavy metals (Garbisu and Alkorta 2002).

Hyperaccumulators

Particular interest is aroused in plants that can store remarkably large amounts of heavy elements, far in excess of the levels found in other species without showing any symptoms of phytotoxicity known as "hyperaccumulators."

They are characterized by an elevated degree of heavy metal absorption, a rapid translocation from roots to shoots and mechanism of detoxification. Reports revealed that heavy metal concentration accumulated in above-ground organs, especially leaves of hyperaccumulators are about four orders of magnitude higher than those found in non-hyperaccumulating species (Rascio and Navari-Izzo 2011; Sarma 2011). So far, about 720 species have been described as hyperaccumulators with some species exhibiting hyperaccumulation of more than one metal and the number has been constantly rising. The hyperaccumulator species are from 100 families, and among the families most strongly represented are the *Brassicaceae* (about 25% of species) and the *Phyllanthaceae* (Reeves et al. 2018). Using hyperaccumulators may prove to be an easy and inexpensive tool for reducing heavy metals in the environment.

In most hyperaccumulators, the primary methods for metal ion transport are the molecules encoded by genes of the ZIP family (ZRT/IRT-related proteins) which is a unique characteristic of plants containing a family of genes related to coding transporters of variety of cations, including Cd and Zn. Moreover, more families such as the HMA, MATE, YSL and MTP have been described too as being involved in metal transport (Guerinot 2000; Mizuno et al. 2008; Milner et al. 2013). A report by Gao et al. (2013) indicates that a few genes related to alternation of cell wall or metal transport were more induced or expressed at greater levels in hyperaccumulators when compare to non-hyperaccumulators in response to heavy metal exposure. However, despite many advantages of hyperaccumulators, the limitation of using this group of plants is their low growth rate and biomass production. Examples of hyperaccumulators are listed in Table 1.

Table 1. Some important metal hyperaccumulators

Heavy metal	Plant species	Reference
As	*Agrostis castellana*	De Koe 1994
	Agrostis delicatula	De Koe 1994
	Cyanoboletus pulverulentus	Braeuer et al. 2018
	Pteris cretica var. nervosa.	Zang et al. 2017
	Pteris umbrosa R. Br.	Koller et al. 2007
	Pteris vittata	Yan et al. 2019
Cd	*Arabidopsis halleri*	Zhang et al. 2017
	Cardamine hirsuta Linn.	Lin et al. 2014
	Ceratophyllum demersum	Matache et al. 2013
	Galinsoga parviflora	Lin et al. 2014
	Gnaphalium affine D. Don.	Lin et al. 2014
	Lantana camara L.	Liu et al. 2019
	Microsorum pteropus	Lan et al. 2019
	Pterocypsela laciniata	Zhong et al. 2019
	Sedum alfredii Hance	Gao et al. 2013
	Solanum nigrum L.	Li et al., 2019
	Thlaspi caerulescens	van der Pas and Ingle 2019

Heavy metal	Plant species	Reference
Cd	*Viola baoshanensis*	Shu et al. 2019
	Youngia erythrocarpa	Lin et al. 2015
Cr	*Amaranthus dubius*	Mellem et al. 2012
	Brassica juncea	Elektorowicz and Keropian 2015
	Leersia hexandra Swartz	Zhang et al. 2007
	Nopalea cochenillifera	Adki et al. 2013
	Prosopis laevigata	Buendía-González et al. 2010
	Spartina argentinensis	Redondo-Gómez et al. 2011
	Thlaspi caerulescens	Mandáková et al. 2015
Cu	*Aeolanthus biformifolius*	Malaisse et al. 1978
	Calandula officinalis L.	Goswami and Das 2016
	Centella asiatica	Mokhtar et al. 2011
	Eichhornia crassipes	Mokhtar et al. 2011
	Pistia stratiotes	Čadková et al. 2015
	Prosopis pubescens	Zappala et al. 2013
	Salvinia cucullata	Das and Goswami 2017
Pb	*Brassica juncea*	Lim et al. 2004
	Scirpus grossus	Tangahu et al. 2013
	Sedum alfredii Hance	Gao et al. 2013
	Sesbania drummondii	Sahi et al. 2002
Hg	*Dittrichia viscosa*	Shehu et al. 2014
	Medicago sativa L.	Shehu et al. 2014
	Mentha arvensis	Manikandan et al. 2015
	Sesbania drummondii	Venkatachalam et al. 2009
Ni	*Alyssum bertolonii*	Mengoni et al. 2012
	Alyssum discolor	Bayramoglu et al. 2012
	Phyllanthus rufuschaneyi	van der Pas and Ingle 2019
	Pycnandra acuminata	Jaffré et al. 2018
	Senecio conrathii	van der Pas and Ingle 2019
	Streptanthus polygaloides	van der Pas and Ingle 2019
Zn	*Anthyllis vulneraria* L.	van der Pas and Ingle 2019
	Arabidopsis halleri	Zhang et al. 2017
	Thlaspi caerulescens	van der Pas and Ingle 2019
	Sedum alfredii Hance	Gao et al. 2013

Genetically Modified Plants

Recent accomplishments in genetic engineering of plants might be the key to amelioration of phytoremediation. The most prominent approach is to introduce and overexpress genes associated with the plant's ability to tolerate, accumulate, stabilize and mobilize heavy metals. Moreover, attention has focused mostly on reduction of time required for clean-up and the obtaining plants characterized by increased growth rate and enlarged plant biomass production. Several genes involved in primary metabolism, encoding metal-binding proteins and metal transporters, enzymes for the biosynthesis of metal ligands, and detoxification of metals by chemical modification have been used as targets to generate transgenic plants (Fasani et al. 2018). Many studies demonstrated and confirmed that genetically modified plants including the hyperaccumulators – *Brassica juncea, Helianthus annuus, Lycopersicon esculentum and Liriodendron tulipifera*, have increased uptake, translocation, toleration and storage of Cd, Zn, Pb, Ni in the shoots (Kotrba et al. 2009; Sarwar et al. 2017).

TYPES OF PHYTOREMEDIATION TECHNOLOGY

Phytoextraction

Phytoextraction (also called phytoaccumulation, phyto-mining or bio-mining) is based on removal of heavy metals due to accumulation in the above-ground parts of plants. Plants used for this method are characterized by rapid growth and production of biomass, tolerance to conditions prevailing on various types of post-industrial dumping grounds, i.e., the presence of increased levels of toxic metals in the substrate, drought, nutrient deficiency, high temperatures and their variations, and a variety of soil pH. Likewise, it is important that plants selected for phytoextraction are able to expand an extensive root system to take up heavy metals such as Pb, Cd, Ni, Cu, and Cr from broader soil areas (Robinson et al. 2015;

Suman et al. 2018). Through phytoextraction, harmful elements can be removed from contaminated soil and water.

Salt et al. (1998) distinguished two types of phytoextraction: chelate-assisted or induced phytoextraction and continuous phytoextraction. The first strategies of phytoextraction is based on the fact that the application of metal chelates to the soil significantly enhances metal accumulation by plants while the second one is determined by the common feature of some vegetable organisms to store, translocate and resist elevated quantities of metals over the growth period, especially using hyperaccumulators. The use of non-accumulator plants in the first approach includes either high biomass plants or rapid-growing trees that can be easily cultivated using agricultural methods. It seems that a large amount of biomass might compensate for a comparably low capacity for heavy metal storage. On the other hand, hyperaccumulator plant species have been demonstrated to be potentially helpful in soil remediation, as they can absorb large amounts of heavy metal from polluted environments, however, their low annual biomass accretion tends to limit their usability (Mleczek et al. 2017).

Currently, it is known that metal uptake potential is increased significantly by the application of chelators such as ethylenediaminetetraacetic acid, citric acid, tartaric acid as well as glycine and histidine, nevertheless, the concentration of these elements in edible parts of vegetable organisms is also increased (Chaturvedi et al. 2019). The gathered results highlight the meaningful role of specialist studies in estimating the response of plants growing in massively polluted soils (Budzyńska et al. 2019). Various species of plants such as *Salix x fragilis* L., *Malva verticillata* L., *Sorghum bicolor* L. *Helianthus tuberosus* L., *Phalaris arundinacea* L., *Miscanthus sinensis* Andersson (Mayerová et al. 2017), *Miscanthus x giganteus*, *Sida hermaphrodita* (Kocoń and Jurga 2017), *Acer platanoides* (Budzyńska et al. 2019) and *Amaranthus hypochondriacus* (Tai et al. 2018) are the most useful in the process of phytoextraction. The use of these plant species allowed for successful removal of elements such as As, Cd, Zn and Pb from polluted mining sludge and wastewater–irrigated soil.

Nowadays, this technology is considered as one of the cheapest and environmentally and ecologically-friendly technology known as "green technology" for cleaning an habitat contaminated by heavy metals without adversely affecting the properties of soil (Garbisu and Alkorta 2002). Unfortunately, this is a fairly long-lasting method of phytoremediation limited by slow growth of plants and low biomass efficiency.

Phytostabilization

Phytostabilization (also known as phytoimmobilization) is based on using plants to reduce the mobility of heavy metals in soil. This approach also decreases heavy element bioavailability by precipitation of compounds characterized by lower solubility. Immobilization of contaminants in the ground is based on the following processes: the up taking and storage in the roots, adherence on the surface of roots or transformation these elements into sparingly soluble compounds in the rhizosphere.

Plants suitable for phytostabilization should be native, drought-, salt- and metal-tolerant to decrease water and soil pollution. Moreover, they should not store heavy metals in their above-ground parts, thus preventing their entry to subsequent elements of the food chain, and should be characterized by rapid growth and a dense root system to cover a polluted area in a short period of time (Alkorta et al. 2011; Bolan et al. 2011). Thanks to root secretions and the release of carbon dioxide by roots, the root zone is area in which heavy metals are retained and precipitated (Boisson et al. 2016; Radziemska et al. 2017). Among the most useful plants in this method are *Platycladus orientalis* (Zeng et al. 2018), *Bruguiera cylindrica* (L.) Blume (Sruthi and Puthur 2019), *Prosopis velutina, Acacia farnesiana, Brickellia coulteri, Baccharis sarothroides, Gnaphalium leucocephalum* (Santos et al. 2017), *Festuca rubra* L. (Radziemska et al. 2017), *Alternanthera philoxeroides, Bidens frondosa, Artemisia princeps, Erigeron canadensis, Cynodon dactylon, Setaria plicata, Digitaria sanguinalis, Bidens pilosa* (Yang et al. 2014). To enhance the physical and biological features of polluted ground, natural and synthetic amendments were added at the time of phytostabilization, so this method is known as aided phytostabilization or chemophytostabilization (Garaiyurrebaso et al. 2017). Alteration of the pH, intensifying organic matter content by adding compost and necessary growth compounds, expanding water holding capacity, and decreasing heavy metal bioavailability stimulate phytostabilization (Parmar and Singh 2015). The main problem of this approach is the fact that this method does not solve the problem of pollution definitively, the pollution is only immobilized.

Phytodegradation

Phytodegradation (also known as phytotransformation) is based on decomposition of substances by plants and their associated microorganisms by uptake and transformation of toxic compounds with endogenous enzymes into non-toxic forms. The degradation products generated in this process are embedded into new tissues. These compounds can also be evaporated by stomata and also decomposed into CO_2 and H_2O (Ashraf et al. 2019). Results show that plants take up, translocate, transform and degrade many chemical substances via involvement of cytochrome P450, transferases, and ability to further store substances in vacuoles or in the cell wall (He et al. 2017). Some plants are able to remediate polluted soil, sludge, sediment and ground and surface water by releasing enzymes. Many species of plants such as *Brassica campestris, Festuca arundinacea,* and *Helianthus annuus* have been successfully used in phytodegradation of soil simultaneously contaminated with Cu, Cd, Pb and Ni (Park et al. 2011).

Phytodegradation is used in the cleaning of areas contaminated with not only heavy metals, but also petroleum derivatives, chlorinated derivatives, explosives, drugs, herbicides and even hazardous and common pollutants such as bisphenol A (Newman and Reynolds, 2004; Zazouli et al., 2014; Al-Baldawi, 2018).

Phytostimulation

Phytostimulation (also known as rhizodegradation) is based on plant support of naturally occurring microbial degradation processes by releasing compounds which enhance microbial activity in the rhizosphere (Ashraf et al. 2019). Vegetable organism secretions such as carbohydrates, carboxylic and amino acids can facilitate entire rhizosphere microbial populations such that the overall extended activity will result in increased biodegradation of some pollutants into harmless forms (Dzantor and Kudjo 2007). *Arabidopsis thaliana* and *Brassica campestris* L. are used in phytostimulation (Zahoor et al. 2017; Zhang et al. 2018).

Phytovolatilization

Phytovolatilization is based on transferring pollutants into a volatile state through stomata to the atmosphere. This method found application in the purification of contaminated environments containing selenium, Hg and As (Wang et al. 2012). Conversion of heavy metals into gaseous/volatilized derivatives inside plants requires specific mechanisms involving enzymes or genes. However, despite the relatively high attractiveness of the method, it is the most controversial of the phytoremediation methods and may be a threat to the environment and human health connected with the fact that pollutants are transferred from the soil or waters to the atmosphere where it can be spread (Ali et al. 2013; Limmer and Burken 2016). *Arabidopsis thaliana, Brassica juncea, Pteris vittata* and *Chara canescens* are capable to volatilize selenium and As (Khalid et al. 2016).

Phytofiltration

Phytofiltration is based on using the ability of plant roots or seedlings or excised plant shoots (respectively - rhizofiltration, blastofiltration, caulofiltration) to purify aqueous solutions including waste water, surface water of extracted ground water. Rizophilization is a technology that uses the ability of plant roots to absorb pollutants from aqueous solutions, e.g., from contaminated surface waters. The features of plants useful for this method are: rapid root growth, limited ability to transport contaminants collected to the shoots, high biomass production and tolerance to toxic compounds. Blastofiltration is a method that uses the ability of roots of seedlings growing in aquatic cultures to absorb toxic metals. It is a more effective method than rhizofiltration and can be used as a complementary method for the purification of aqueous solutions (Parmar and Singh 2015). Willows (*Salix* spp.), various aquatic plants such as *Brassica juncea, Helianthus annuus,* and *Nicotiana tabacum* are regarded to be of the more interesting plants with their high tolerance for phytofiltration of soil and water contaminated with trace metals including Cd, Cr, Cu, Ni, Pb, and Zn (Islam et al. 2015; Parmar and Singh 2015; Olguín et al. 2017; Yang et al. 2018).

CONCLUSION

Phytoremediation is proposed as a promising green alternative to competitive method in relation to many traditional methods and can also be used not only as a separate method but also as a complement to conventional remediation methods. One of the advantages of using plants to clean the environment is low cost. Phytoremediation does not require specialized equipment, and techniques are simple. It is an effective form of purifying not only small but also extensive contaminated areas. The use of plants is more effective than conventional methods, because the roots have the ability to penetrate large areas of the rhizosphere and can actively and selectively take up metal ions. This "green technology" can be applied to clean up the contaminated ground without any damaging consequences.

Although biological strategies of removal heavy metals are a promising approach for remediation of polluted areas, they also have some limitations.

One of them is the limited range of purification to the rhizosphere, depending on the root structure. In addition, it is a long-lasting purification process, lasting from a few to several years. Phytoremediation has its limitations also in the environment itself, resulting from the presence of too high concentrations of metals in the soil, toxic even to tolerant populations of plants what suggests that it is appropriate to locations with low to moderate contamination level.

ACKNOWLEDGMENTS

Dorota Bartusik-Aebisher acknowledges support from the National Center of Science NCN (New drug delivery systems-MRI study, Grant OPUS-13 number 2017/25/B/ST4/02481).

REFERENCES

Adki VS, Jadhav JP, Bapat VA (2013) Nopalea cochenillifera, a potential chromium (VI) hyperaccumulator plant. *Environ Sci Pollut Res Int* 20:1173–1180. doi:10.1007/s11356-012-1125-4.

Al–Baldawi IA (2018) Removal of 1,2–Dichloroethane from real industrial wastewater using a sub–surface batch system with Typha angustifolia L. *Ecotoxicol Environ Saf* 147:260–265. doi:10.1016/j.ecoenv.2017.08.022.

AL-Farraj AS, Al-Wabel MI (2007) Heavy metals accumulation of some plant species grown on mining. *J Appl Sci* 7:1170–1175.

Ali H, Khan E, Sajad MA (2013) Phytoremediation of heavy metals--concepts and applications. *Chemosphere* 91:869–881. doi:10.1016/j.chemosphere.2013.01.075.

Ali S, Hussain S, Khan R, Mumtaz S, Ashraf N, Andleeb S, Shakir HA, Tahir HM, Khan MKA, Ulhaq M (2019) Renal toxicity of heavy metals (cadmium and mercury) and their amelioration with ascorbic acid in rabbits. *Environ Sci Pollut Res Int* 26:3909–3920. doi:10.1007/s11356-018-3819-8.

Alkmim Filho JF, Germano A, Dibai WLS, Vargas EA, Melo MM (2014) Heavy metals investigation in bovine tissues in Brazil. *Food Sci Technol* 34:110–115. doi:10.1590/S0101-20612014005000013.

Alkorta I, Becerril JM, Garbisu C (2010) Phytostabilization of metal contaminated soils. *Rev Environ Health* 25:135-46.

Andjelkovic M, Buha Djordjevic A, Antonijevic E, Antonijevic B, Stanic M, Kotur–Stevuljevic J, Spasojevic–Kalimanovska V, Jovanovic M, Boricic N, Wallace D, Bulat Z (2019) Toxic effect of acute cadmium and lead exposure in rat blood, liver, and kidney. *Int J Environ Res Public Health* 16. [in press] doi:10.3390/ijerph16020274.

Ashraf S, Ali Q, Zahir ZA, Ashraf S, Asghar HN (2019) Phytoremediation: Environmentally sustainable way for reclamation of heavy metal polluted soils. *Ecotoxicol Environ Saf* 174:714–727. doi:10.1016/j.ecoenv.2019.02.068.

Bayramoglu G, Arica MY, Adiguzel N (2012) Removal of Ni(II) and Cu(II) ions using native and acid treated Ni–hyperaccumulator plant Alyssum discolor from Turkish serpentine soil. *Chemosphere* 89:302–309. doi:10.1016/j.chemosphere.2012.04.042.

Boisson S, Le Stradic S, Collignon J, Séleck M, Malaisse F, Ngoy Shutcha M, Faucon MP, Mahy G (2016) Potential of copper-tolerant grasses to implement phytostabilisation strategies on polluted soils in South D. R. Congo: Poaceae candidates for phytostabilisation. *Environ Sci Pollut Res Int* 23:13693–13705. doi: 10.1007/s11356-015-5442-2.

Bolan NS, Park JH, Robinson B, Naidu R, Huh KY (2011) Phytostabilization: a green approach to contaminant Containment. *Adv Agron* 112:145–204 doi:10.1016/B978-0-12-385538-1.00004-4.

Braeuer S, Goessler W, Kameník J, Konvalinková T, Žigová A, Borovička J (2018) Arsenic hyperaccumulation and speciation in the edible ink stain bolete (Cyanoboletus pulverulentus). *Food Chem* 242:225–231. doi:10.1016/j.foodchem.2017.09.038.

Budzyńska S, Mleczek P, Szostek M, Goliński P, Niedzielski P, Kaniuczak J, Rissmann I, Rymaniak E, Mleczek M (2019) Phytoextraction of arsenic forms in selected tree species growing in As-polluted mining sludge. *J Environ Sci Health A Tox Hazard Subst Environ Eng* 54:933–942. doi:10.1080/10934529.2019.1609322.

Buendía-González L, Orozco-Villafuerte J, Cruz-Sosa F, Barrera-Díaz C, Vernon-Carter E (2010) Prosopis laevigata a potential chromium (VI) and cadmium (II) hyperaccumulator desert plant. *Bioresour Technol* 101:5862–5867. doi:10.1016/j.biortech.2010.03.027.

Čadková Z, Száková J, Miholová D, Horáková B, Kopecký O, Křivská D, Langrová I, Tlustoš P (2015) Bioaccessibility versus bioavailability of essential (Cu, Fe, Mn, and Zn) and toxic (Pb) elements from phyto hyperaccumulator Pistia stratiotes: potential risk of dietary intake. *J Agric Food Chem* 63:2344–2354. doi:10.1021/jf5058099.

Chaturvedi R, Favas P, Pratas J, Varun M, Paul MS (2019) EDTA–Assisted Metal Uptake in Raphanus sativus L. and Brassica oleracea L.: Assessment of Toxicity and Food Safety. *Bull Environ Contam Toxicol.* [in press] doi:10.1007/s00128-019-02651-9.

Chen YY, Tang MY, Wang ST, Wang Q, Zhan WX, Huang G (2016) Heavy metal pollution assessment of farmland soil in China based on bibliometrics. *Chin J Soil Sci* 47:219–225.

Das S, Goswami S (2017) Copper phytoextraction by Salvinia cucullata: biochemical and morphological study. *Environ Sci Pollut Res Int* 24:1363–1371. doi: 10.1007/s11356–016–7830–7.

De Koe T (1994) Agrostis castellana and Agrostis delicatula on heavy metal and arsenic enriched sites in NE Portugal. *Sci Total Environ* 145:103–109.

Dzantor, E. Kudjo (2007–03–01). "Phytoremediation: the state of rhizosphere 'engineering' for accelerated rhizodegradation of xenobiotic contaminants." *Journal of Chemical Technology & Biotechnology.* 82 (3): 228–232.

Elektorowicz M, Keropian Z (2015) Lithium, vanadium and chromium uptake ability of Brassica juncea from Lithium mine tailings. *Int J Phytoremediation* 17:521–528. doi:10.1080/15226514.2013.876966.

Ernst WHO, Verkleji JAC, Schat H (1992) Metal tolerance in plants. *Acta Bot Neerl* 41:229–248.

Farid M, Shakoor MB, Ehsan S, Ali S, Zubair M, Hanif MA (2013) Morphological, physiological and biochemical responses of different plant species to Cd stress. *IJCBS* 3:53–60.

Fasani E, Manara A, Martini F, Furini A, DalCorso G (2018) The potential of genetic engineering of plants for the remediation of soils contaminated with heavy metals. *Plant Cell Environ* 41:1201–1232. doi:10.1111/pce.12963.

Filipović–Trajković R, Ilić ZS, Šunić L, Andjelković S (2012) The potential of different plant species for heavy metals accumulation and distribution. *JFAE* 10:959–964.

Gao J, Sun L, Yang X, Liu JX (2013) Transcriptomic analysis of cadmium stress response in the heavy metal hyperaccumulator Sedum alfredii Hance. *PLoS One* 8:e64643. doi:10.1371/journal.pone.0064643.

Garaiyurrebaso O, Garbisu C, Blanco F, Lanzén A, Martín I, Epelde L, Becerril JM, Jechalke S, Smalla K, Grohmann E, Alkorta I (2017) Long-term effects of aided phytostabilisation on microbial communities of metal-contaminated mine soil. *FEMS Microbiol Ecol* 93:252. doi:10.1093/femsec/fiw252.

Garbisu C, Alkorta I (2001) Phytoextraction: a cost–effective plant–based technology for the removal of metals from the environment. *Bioresour Technol* 77:229–236.

Goswami S, Das S (2016) Copper phytoremediation potential of Calandula officinalis L. and the role of antioxidant enzymes in metal tolerance. *Ecotoxicol Environ Saf* 126:211–218. doi:10.1016/j.ecoenv.2015.12.030.

Guerinot ML (2000) The ZIP family of metal transporters. *Biochim Biophys Acta* 1465:190–198.

He Y, Langenhoff AAM, Sutton NB, Rijnaarts HHM, Blokland MH, Chen F, Huber C, Schröder P (2017) Metabolism of ibuprofen by Phragmites australis: uptake and phytodegradation. *Environ Sci Technol* 51:4576–4584. doi:10.1021/acs.est.7b00458.

Heba H, Abu Zeid I, Abuzinadah OA, Faragalla A, Al–Hasaw Z (2015) Determination of Some Heavy Metals in Tissues and Organs of Three Commercial Fish Species at Al–Hudaydah, Red Sea Coast of

Western Yemen. *WJFMS* 7(3):198–208. doi:10.5829/idosi.wjfms. 2015.7.3.9429.

Islam MS, Saito T, Kurasaki M (2015) Phytofiltration of arsenic and cadmium by using an aquatic plant, Micranthemum umbrosum: phytotoxicity, uptake kinetics, and mechanism. *Ecotoxicol Environ Saf* 112:193–200. doi:10.1016/j.ecoenv.2014.11.006.

Jabeen R, Ahmad A, Iqbal M (2009) Phytoremediation of heavy metals: physiological and molecular mechanisms. *Bot Rev* 75:339–364.

Jaffré T, Reeves RD, Baker AJM, Schat H, van der Ent A (2018) The discovery of nickel hyperaccumulation in the New Caledonian tree Pycnandra acuminata 40 years on: an introduction to a Virtual Issue. *New Phytol* 218:397–400. doi:10.1111/nph.15105.

Keogh JP, Steffen B, Siegers CP (1994) Cytotoxicity of heavy metals in the human small intestinal epithelial cell line I-407: the role of glutathione. *J Toxicol Environ Health* 43:351–359.

Khalid S, Shahid M, Niazi NK, Murtaza B, Bibi I, Dumat C (2016) A comparison of technologies for remediation of heavy metal contaminated soils. *J Geochem Explor* 182: 247–268. doi:10.1016/j.gexplo.2016.11.021ff.

Kocoń A, Jurga B (2017) The evaluation of growth and phytoextraction potential of Miscanthus x giganteus and Sida hermaphrodita on soil contaminated simultaneously with Cd, Cu, Ni, Pb, and Zn. *Environ Sci Pollut Res Int* 24:4990–5000. doi:10.1007/s11356-016-8241-5.

Koller CE, Patrick JW, Rose RJ, Offler CE, MacFarlane GR (2007) Pteris umbrosa R. Br. as an arsenic hyperaccumulator: accumulation, partitioning and comparison with the established As hyperaccumulator Pteris vittata. *Chemosphere* 66:1256–1263.

Kotrba P, Najmanova J, Macek T, Ruml T, Mackova M (2009) Genetically modified plants in phytoremediation of heavy metal and metalloid soil and sediment pollution. *Biotechnol Adv* 27:799–810. doi:10.1016/j.biotechadv.2009.06.003.

Krstic B, Stankovic D, Igic R, Nikolic N (2007) The potential of different plant species for nickel accumulation. *J Biotechnol Equip* 21:431–436.

Lan XY, Yan YY, Yang B, Li XY, Xu FL (2019) Subcellular distribution of cadmium in a novel potential aquatic hyperaccumulator – Microsorum pteropus. *Environ Pollut* 248:1020–1027. doi:10.1016/j.envpol.2019.01.123.

Li K, Yang B, Wang H, Xu X, Gao Y, Zhu Y (2019) Dual effects of biochar and hyperaccumulator Solanum nigrum L. on the remediation of Cd–contaminated soil. *PeerJ* 7:e6631. doi:10.7717/peerj.6631.

Li Q, Liu H, Alattar M, Jiang S, Han J, Ma Y, Jiang C (2015) The preferential accumulation of heavy metals in different tissues following frequent respiratory exposure to PM2.5 in rats. *Sci Rep* 5:16936. doi:10.1038/srep16936.

Lim JM, Salido AL, Butcher BJ (2004) Phytoremediation of lead using Indian mustard (Brassica juncea) with EDTA and electrodics. *Microchem J* 76:3–9.

Limmer M, Burken J (2016) Phytovolatilization of Organic Contaminants. *Environ Sci Technol* 50(13):6632–6643. doi:10.1021/acs.est.5b04113.

Lin L, Jin Q, Liu Y, Ning B, Liao M, Luo L (2014) Screening of a new cadmium hyperaccumulator, Galinsoga parviflora, from winter farmland weeds using the artificially high soil cadmium concentration method. *Environ Toxicol Chem* 33:2422–2428. doi:10.1002/etc.2694.

Lin L, Ning B, Liao M, Ren Y, Wang Z, Liu Y, Cheng J, Luo L (2015) Youngia erythrocarpa, a newly discovered cadmium hyperaccumulator plant. *Environ Monit Assess* 187:4205. doi:10.1007/s10661-014-4205-8.

Lin L, Shi J, Liu Q, Liao M, Mei L (2014) Cadmium accumulation characteristics of the winter farmland weeds Cardamine hirsuta Linn. and Gnaphalium affine D. Don. *Environ Monit Assess* 186:4051–4506. doi:10.1007/s10661–014–3679–8.

Liu S, Ali S, Yang R, Tao J, Ren B (2019) A newly discovered Cd–hyperaccumulator Lantana camara L. *J Hazard Mater* 371:233–242. doi:10.1016/j.jhazmat.2019.03.016.

Malaisse F, Gregoire J, Brooks RR, Morrison RS, Reeves RD (1978) Aeolanthus biformifolius De Wild.: A hyperaccumulator of copper from Zaire. *Science* 199:887–888.

Mandáková T, Singh V1, Krämer U1, Lysak MA (2015) Genome structure of the heavy metal hyperaccumulator Noccaea caerulescens and its stability on metalliferous and nonmetalliferous soils. *Plant Physiol* 169:674–689. doi:10.1104/pp.15.00619.

Manikandan R, Sahi SV, Venkatachalam P (2015) Impact assessment of mercury accumulation and biochemical and molecular response of Mentha arvensis: a potential hyperaccumulator plant. *Sci World J* 2015:715217. doi:10.1155/2015/715217.

Marth E, Jelovcan S, Kleinhappl B, Gutschi A, Barth S (2001) The effect of heavy metals on the immune system at low concentrations. *Int J Occup Med Environ Health* 14:375–386.

Matache ML, Marin C, Rozylowicz L, Tudorache A (2013) Plants accumulating heavy metals in the Danube River wetlands. *J Environ Health Sci Eng* 11:39. doi:10.1186/2052-336X-11-39.

Matović V, Buha A, Dukić–Ćosić D, Bulat Z (2015) Insight into the oxidative stress induced by lead and/or cadmium in blood, liver and kidneys. *Food Chem Toxicol* 78:130–140. doi:10.1016/j.fct.2015.02.011.

Mayerová M, Petrová Š, Madaras M, Lipavský J, Šimon T, Vaněk T (2017) Non–enhanced phytoextraction of cadmium, zinc, and lead by high–yielding crops. *Environ Sci Pollut Res Int* 24:14706–14716. doi:10.1007/s11356-017-9051-0.

Mellem JJ, Baijnath H, Odhav B (2012) Bioaccumulation of Cr, Hg, As, Pb, Cu and Ni with the ability for hyperaccumulation by Amaranthus dubius. *Afr J Agric Res* 7:591–596. doi:10.5897/AJAR11.1486.

Mengoni A, Cecchi L, Gonnelli C (2012) Nickel hyperaccumulating plants and Alyssum bertolonii: Model systems for studying biogeochemical interactions in serpentine soils. In: Kothe E., Varma A. (eds) Bio–Geo Interactions in Metal–Contaminated Soils. Soil Biology, Springer, Berlin, Heidelberg.

Milner MJ, Seamon J, Craft E, Kochian LV (2013) Transport properties of members of the ZIP family in plants and their role in Zn and Mn homeostasis. *J Exp Bot* 64:369–381. doi:10.1093/jxb/ers315.

Mizuno T, Hirano K, Kato S, Obata H (2008) Cloning of ZIP family metal transporter genes from the manganese hyperaccumulator plant Chengiopanax sciadophylloides and its metal transport and resistance abilities in yeast. *Soil Sci Plant Nutr* 54:86–94.

Mleczek M, Goliński P, Krzesłowska M, Gąsecka M, Magdziak Z, Rutkowski P, Budzyńska S, Waliszewska B, Kozubik T, Karolewski Z, Niedzielski P (2017) Phytoextraction of potentially toxic elements by six tree species growing on hazardous mining sludge. *Environ Sci Pollut Res Int* 24:22183–22195. doi:10.1007/s11356-017-9842-3.

Mokhtar H, Morad N, Fizri FF (2011) Hyperaccumulation of copper by two species of aquatic plants. *IPCBEE* 8:115–118.

Morkunas I, Woźniak A, Mai VC, Rucińska–Sobkowiak R, Jeandet P (2018) The role of heavy metals in plant response to biotic stress. *Molecules* 23:2320. doi: 10.3390/molecules23092320.

Newman LA, Reynolds CM (2004) Phytodegradation of organic compounds. *Curr Opin Biotechnol* 15:225–230.

Olguín EJ, García–López DA, González–Portela RE, Sánchez–Galván G (2017) Year–round phytofiltration lagoon assessment using Pistia stratiotes within a pilot–plant scale biorefinery. *Sci Total Environ* 592:326–333. doi:10.1016/j.scitotenv.2017.03.067.

Pamphlett R, Bishop DP, Kum Jew S, Doble PA (2018) Age–related accumulation of toxic metals in the human locus ceruleus. *PLoS One* 13:e0203627. doi:10.1371/journal.pone.0203627.

Park S, Kim KS, Kim JT, Kang D, Sung K (2011) Effects of humic acid on phytodegradation of petroleum hydrocarbons in soil simultaneously contaminated with heavy metals. *J Environ Sci* (China) 23:2034–2041.

Park YH, Kim D, Dai J, Zhang Z (2015) Human bronchial epithelial BEAS–2B cells, an appropriate in vitro model to study heavy metals induced carcinogenesis. *Toxicol Appl Pharmacol* 287:240–245. doi:10.1016/j.taap.2015.06.008.

Parmar S, Singh V (2015) Phytoremediation approaches for heavy metal pollution: a review. *J Plant Sci Res* 2:139.

Radziemska M, Vaverková MD, Baryła A (2017) Phytostabilization– management strategy for stabilizing trace elements in contaminated soils. *Int J Environ Res Public Health* 14:958. doi:10.3390/ijerph 14090958.

Rascio N, Navari–Izzo F (2011) Heavy metal hyperaccumulating plants: how and why do they do it? And what makes them so interesting? *Plant Sci* 180:169–181. doi:10.1016/j.plantsci.2010.08.016.

Redondo-Gómez S, Mateos-Naranjo E, Vecino-Bueno I, Feldman SR (2011) Accumulation and tolerance characteristics of chromium in a cordgrass Cr–hyperaccumulator, Spartina argentinensis. *J Hazard Mater* 185:862–869. doi:10.1016/j.jhazmat.2010.09.101.

Reeves RD, Baker AJM, Jaffré T, Erskine PD, Echevarria G, van der Ent A (2018) A global database for plants that hyperaccumulate metal and metalloid trace elements. *New Phytol* 218:407–411. doi:10.1111/nph.14907.

Rehman K, Fatima F, Waheed I, Akash MSH (2018) Prevalence of exposure of heavy metals and their impact on health consequences. *J Cell Biochem* 119:157–184. doi:10.1002/jcb.26234.

Robinson BH, Anderson CWN, Dickinson NM (2015) Phytoextraction: Where's the action? *J Geochem Explor* 151:34–40. doi:10.1016/j.gexplo.2015.01.001.

Rudy M, Znamirowska A, Zin M (2007) Level of accumulation of selected heavy metals in horse tissue as a function of age. *Medycyna Wet* 63:1303–1306.

Sahi SV, Bryant NL, Sharma NC, Singh SR (2002) Characterization of a lead hyperaccumulator shrub, Sesbania drummondii. *Environ Sci Technol* 36:4676–4680.

Salt DE, Blaylock M, Kumar NPBA, Dushenkov·V, Ensley D, Chet I, Raskin I (1995) Phytoremediation: a novel strategy for the removal of toxic metals from the environment using plants. *Biotechnology* 13:468–474.

Salt DE, Smith RD, Raskin I (1998) Phytoremediation. *Annu Rev Plant Physiol Plant Mol Biol* 49:643–668.

Santos AE, Cruz-Ortega R, Meza-igueroa D, Romero FM, Sanchez-Escalante JJ, Maier RM, Neilson JW, Alcaraz LD, Molina Freaner FE (2017) Plants from the abandoned Nacozari mine tailings: evaluation of their phytostabilization potential. *PeerJ* 5:e3280. doi:10.7717/peerj.3280.

Sarma H (2011) Metal hyperaccumulation in plants: a review focusing on phytoremediation technology. *Environ Sci Technol* 4:118–138. doi:10.3923/jest.2011.118.138.

Sas-Nowosielska H, Pawlas N (2015) Heavy metals in the cell nucleus – role in pathogenesis. *Acta Biochim Pol* 62:7–13.

Sarwar N, Imran M, Shaheen MR, Ishaque W, Kamran MA, Matloob A, Rehim A, Hussain S (2017) Phytoremediation strategies for soils contaminated with heavy metals: Modifications and future perspectives. *Chemosphere* 171:710–721. doi:10.1016/j.chemosphere.2016.12.116.

Shahid M, Pourrut B, Dumat C, Nadeem M, Aslam M, Pinelli E (2014) Heavy–metal–induced reactive oxygen species: phytotoxicity and physicochemical changes in plants. *Rev Environ Contam Toxicol* 232:1–44. doi:10.1007/978–3–319–06746–9_1.

Shallari S, Schwartz C, Hasko A, Morel JL (1997) Heavy metals in soils and plants of serpentine and industrial site of Albania. *Sci Total Environ* 209:133–142.

Shehu J, Imeri A, Kupe L, Dodona E, Shehu A, Mullaj A (2014) Hyperaccumulators of mercury in the industrial area of a PVC factory in Vlora (Albania). *Arch Biol Sci* 6:1457–1464. doi:10.2298/ABS1404457S.

Shu H, Zhang J, Liu F, Bian C, Liang J, Liang J, Liang W, Lin Z, Shu W, Li J, Shi Q, Liao B (2019) Comparative Transcriptomic studies on a cadmium hyperaccumulator Viola baoshanensis and its non–tolerant counterpart V. inconspicua. *Int J Mol Sci* 20:1906. doi:10.3390/ijms20081906.

Sruthi P, Puthur JT (2019) Characterization of physiochemical and anatomical features associated with enhanced phytostabilization of copper in Bruguiera cylindrica (L.) Blume. *Int J Phytoremediation* 21:1–19. doi:10.1080/15226514.2019.1633263.

Suman J, Uhlik O, Viktorova J, Macek T (2018) Phytoextraction of heavy metals: a promising tool for clean–up of polluted environment? *Front Plant Sci* 9:1476. doi:10.3389/fpls.2018.01476.

Tai Y, Yang Y, Li Z, Yang Y, Wang J, Zhuang P, Zou B (2018) Phytoextraction of 55–year–old wastewater–irrigated soil in a Zn–Pb mine district: effect of plant species and chelators. *Environ Technol* 39:2138–2150. doi:10.1080/09593330.2017.1351493.

Takahashi R, Bashir K, Ishimaru Y, Nishizawa NK, Nakanishi H (2012) The role of heavy–metal ATPases, HMAs, in zinc and cadmium transport in rice. *Plant Signal Behav* 7:1605–1607. doi:10.4161/psb.22454.

Tangahu BV, Abdullah SR, Basri H, Idris M, Anuar N, Mukhlisin M (2013) Phytotoxicity of wastewater containing lead (Pb) effects Scirpus grossus. *Int J Phytoremediation* 15:814–826. doi:10.1080/15226514.2012.736437.

Tangahu BV, Sheikh Abdullah SR, Basri H, Idris M, Anuar N, Mukhlisin M (2011) A review on heavy metals (As, Pb, and Hg) uptake by plants through phytoremediation. *Int J Chem Eng* 2011:939161. doi.10.1155/2011/939161.

Timchalk C, Lin Y, Weitz KK, Wu H, Gies RA, Moore DA, Yantasee W (2006) Disposition of lead (Pb) in saliva and blood of Sprague–Dawley rats following a single or repeated oral exposure to Pb–acetate. *Toxicology* 222:86–94. doi:10.1016/j.tox.2006.01.030.

van der Pas L, Ingle RA (2019) Towards an Understanding of the molecular basis of nickel hyperaccumulation in plants. *Plants* (Basel) 8:11. doi:10.3390/plants8010011.

Vella V, Malaguarnera R, Lappano R, Maggiolini M, Belfiore A (2017) Recent views of heavy metals as possible risk factors and potential preventive and therapeutic agents in prostate cancer. *Mol Cell Endocrinol* 457:57–72. doi:10.1016/j.mce.2016.10.020.

Venkatachalam P, Srivastava AK, Raghothama KG, Sahi SV (2009) Genes induced in response to mercury–ion–exposure in heavy metal hyperaccumulator Sesbania drummondii. *Environ Sci Technol* 43:843–850.

Wang J, Feng X, Anderson CW, Xing Y, Shang L (2012) Remediation of mercury contaminated sites – A review. *J Hazard Mater* 221–222:1–18. doi:10.1016/j.jhazmat.2012.04.035.

Wuana RA, Okieimen FE (2011) Heavy metals in contaminated soils: a review of sources, chemistry, risks and best available strategies for remediation. *ISRN Ecology* 2011:402647. doi:10.5402/2011/402647.

Yan H, Gao Y, Wu L, Wang L, Zhang T, Dai C, Xu W, Feng L, Ma M, Zhu YG, He Z. (2019) Potential use of the Pteris vittata arsenic hyperaccumulation–regulation network for phytoremediation. *J Hazard Mater* 368:386–396. doi:10.1016/j.jhazmat.2019.01.072.

Yang Q, Li Z, Lu X, Duan Q, Huang L, Bi J (2018) A review of soil heavy metal pollution from industrial and agricultural regions in China: Pollution and risk assessment. *Sci Total Environ* 642:690–700. doi:10.1016/j.scitotenv.2018.06.068.

Yang S, Liang S, Yi L, Xu B, Cao J, Guo Y, Zhou Y (2014) Heavy metal accumulation and phytostabilization potential of dominant plant species growing on manganese mine tailings. *Front Env Sci Eng* 8:394–404. doi:10.1007/s11783-013-0602-4.

Yang W, Zhao F, Ding Z, Shohag MJI, Wang Y, Zhang X, Zhu Z, Yang X (2018) Screening of 19 Salix clones in effective phytofiltration potentials of manganese, zinc and copper in pilot–scale wetlands. *Int J Phytoremediation* 20:1275–1283. doi:10.1080/15226514.2014.898020.

Zahoor M, Irshad M, Rahman H, Qasim M, Afridi SG, Qadir M, Hussain A (2017) Alleviation of heavy metal toxicity and phytostimulation of Brassica campestris L. by endophytic Mucor sp. MHR–7. *Ecotoxicol Environ Saf* 142:139–149. doi:10.1016/j.ecoenv.2017.04.005.

Zappala MN, Ellzey JT, Bader J, Peralta–Videa JR, Gardea–Torresdey J (2013) Prosopis pubescens (screw bean mesquite) seedlings are hyperaccumulators of copper. *Arch Environ Contam Toxicol* 65:212–223. doi:10.1007/s00244-013-9904-6.

Zazouli MA, Mahdavi Y, Bazrafshan E, Balarak D (2014) Phytodegradation potential of bisphenolA from aqueous solution by Azolla Filiculoides. *J Environ Health Sci Eng* 12:66. doi:10.1186/2052-336X-12-66.

Zeng P, Guo Z, Xiao X, Cao X, Peng C (2018) Response to cadmium and phytostabilization potential of Platycladus orientalis in contaminated soil. *Int J Phytoremediation* 20:1337–1345. doi:10.1080/15226514. 2018.1501338.

Zhang X, Li X, Yang H, Cui Z (2018) Biochemical mechanism of phytoremediation process of lead and cadmium pollution with Mucor circinelloides and Trichoderma asperellum. *Ecotoxicol Environ Saf* 157:21–28. doi:10.1016/j.ecoenv.2018.03.047.

Zhang X, Yang X, Wang H, Li Q, Wang H, Li Y (2017) A significant positive correlation between endogenous trans–zeatin content and total arsenic in arsenic hyperaccumulator Pteris cretica var. nervosa. *Ecotoxicol Environ Saf* 138:199–205. doi:10.1016/j.ecoenv. 2016.12.031.

Zhang XH, Liu J, Huang HT, Chen J, Zhu NY, Wang DK (2007) Chromium accumulation by the hyperaccumulator plant Leersia hexandra Swartz. *Chemosphere* 67:1138–1143.

Zhang Z, Wen X, Huang Y, Inoue C, Liang Y (2017) Higher accumulation capacity of cadmium than zinc by Arabidopsis halleri ssp. germmifera in the field using different sowing strategies. *Plant Soil* 418:165–176. doi:10.1007/s11104-017-3285-y.

Zhong L, Lin L, Liao M, Wang J, Tang Y, Sun G, Liang D, Xia H, Wang X, Zhang H, Ren W (2019) Phytoremediation potential of Pterocypsela laciniata as a cadmium hyperaccumulator. *Environ Sci Pollut Res Int* 26:13311–13319. doi:10.1007/s11356–019–04702–4.

Zhou Q, Gu Y, Yue X, Mao G, Wang Y, Su H, Xu J, Shi H, Zou B, Zhao J, Wang R (2017) Combined toxicity and underlying mechanisms of a mixture of eight heavy metals. *Mol Med Rep* 15:859–866. doi:10.3892/mmr.2016.6089.

In: Environmental Science of Heavy Metals   ISBN: 978-1-53617-831-9
Editor: Dorota Bartusik-Aebisher   © 2020 Nova Science Publishers, Inc.

Chapter 5

ANALYTICAL METHODS FOR THE DETECTION AND DETERMINATION OF HEAVY METALS IN WATER

Rafał Podgórski, Dominika Podgórska, David Aebisher, Sabina Galiniak, Tomasz Kubrak and Dorota Bartusik-Aebisher[*]

Faculty of Medicine, University of Rzeszow, Poland

ABSTRACT

Activities connected with urbanization and industrialization processes such as extensive farming, mining and chemical industry has caused release of toxic heavy metals into natural groundwater, oceans, seas, lakes, rivers, and soils. In certain amounts, several heavy metal ions are required for plant metabolism as essential micronutrients, however, they might become extremely harmful when they accumulate at high concentration in natural environments. Furthermore, they aren't

---

[*] Corresponding Author's Email: dbartusik-aebisher@ur.edu.pl.

biodegradable and remain in the environment for very long time. Herein are presented analytical methods used to assay levels of heavy metals in water. Heavy metal ions can be analyzed by various methods, with the choice often depending on required sensitivity and precision.

The most suitable analytical methods to determine heavy metals in water are: atomic absorption spectrometry, electrochemical methods, colorimetric and chromatographic techniques. Atomic absorption spectrometry is a very sensitive and selective technique, but it requires quite expensive equipment, use of complicated and painstaking operational procedures, and detection time is long. The advantages of electrochemical techniques include: simplicity, portability, rapid, sensitive and low-cost analysis. High performance liquid chromatography hyphenated to inductively coupled plasma mass spectrometry is currently one of the most powerful tolls for the purpose of analysis heavy metals ions in different matrices including water.

Keywords: heavy metals analysis, heavy metal toxicity, atomic absorption spectrometry, electrochemical methods

INTRODUCTION

Heavy metals such as mercury (Hg), lead (Pb), cadmium (Cd), zinc (Zn) and chromium (Cr) generally are loosely defined as the elements for which density is higher than 5 g/cm^3 and have an adverse effect on the natural environment (Järup 2003). They occur naturally in the earth's crust but their geochemical cycles has diametrically changed during last decades due to extensive industrial and urbanization activities. From a total of about 90 naturally occurring elements, approximately 50 (mostly d-block) are considered to be heavy metals. This term is also transposable to the naturally occurring p-block elements as well as to actinide and lanthanide elements (Nies 1999). Moreover, metalloids such as arsenic (As) are often included in the group of heavy metals due to similarities in chemical properties and environmental behavior (Chen et al. 1999). Heavy metal ions are water soluble and can be assimilated by living organisms. Once they enter the food chain, large concentrations of heavy metals may accumulate in tissue causing various harmful effects on human health

(Kurniawan et al. 2006; Rehman et al. 2018). Organisms in the environment may not be able to excrete all heavy metal ions consumed leading to accumulation. Some of them, such as Zn, Cu, Fe, Mn, and Co are required by the human body, all are toxic if consumed in high enough concentration. The heavy metals Pb and Hg have no known positive effects on human health (Chronopoulos et al. 1997).

Heavy Metal Toxicity

Heavy metal toxicity depends on several factors including the dose, type of metal, route of exposure as well as the age, gender, genetics, and nutritional status of exposed individuals.

They can accumulate in various organs such as the heart, kidney, liver or brain and disturb metabolism in diverse ways. The pattern of metals interacting with cellular macromolecules, metabolic and signal transduction pathways and genetic processes is very complex (Beyersmann and Hartwig 2008). Many metallic compounds undergo biotransformation through metabolic processes such as alkylation or reduction to lower oxidation states and can be carcinogenic and mutagenic (Genestra 2007). Exposure to heavy metals occurs through different routes such as: oral (ingestion contaminated water and/or food), inhalation or dermal contact and even by the parenteral route. The most common source of heavy metal exposure is contaminated drinking water which results in various health issues including neuronal and cardiovascular disorders, renal damage, increased risk of diabetes and cancer and ultimately higher rates of morbidity and mortality. (Rusyniak et al. 2010; Hyder et al. 2013).

Chronic exposure to cadmium can have adverse effects such as lung cancer, pulmonary adenocarcinomas, prostatic proliferative lesions, bone fractures, kidney dysfunction, and hypertension (Orlowski and Piotrowski 2003; Huat et al. 2019). Persistent exposure to arsenic may cause neurologic and neurobehavioral disorders, diabetes, cardiovascular and peripheral vascular disease, hematologic disorders and skin cancer (Singh et al. 2007; Żukowska and Biziuk 2008; Dong et al. 2011; Baig et al.

2012). The mechanism of lead toxicity is connected mostly to the ability of Pb^{2+} to exchange with other bivalent cations such as Mg^{2+}, Ca^{2+}, Fe^{2+} and with monovalent ions such as Na^+. Excessive intake of the Pb can impair many of the body's systems such as: the hematopoietic, reproductive, endocrine, skeletal, nervous, cardiovascular and immune systems (Flora et al. 2008; Zhang et al. 2012). Nickel is one of the most common allergens in the world (Torres et al. 2009; Zambelli and Ciurli 2013; Zambelli et al. 2016). Persistent nickel exposure might cause cardiovascular and respiratory systems diseases, or nephropathy. Nickel causes the formation of reactive oxygen species like as hydrogen peroxide (H_2O_2), superoxide radical ($O_2^{\cdot -}$) and hypochlorous acid (HOCl), in neutrophils and monocytes (Guo et al. 2016). Ultimately, it is well documented that nickel has teratogenic and carcinogenic potential also ((France) International Agency for Research on Cancer et al. 1990). Because of good lipid solubility, Hg can easily cross both the placental and blood–brain barrier. That makes the brain, kidney and liver also the most liable and poisoned organs in human body after Hg exposure (Park and Zheng 2012; Rahman and Singh 2019). Intensive, acute exposure to Hg vapor causes severe pneumonitis and acute necrotizing bronchitis. Sustained exposure induces erethism, mercurial tremor and immune system dysfunction (Bernhoft 2012; Markiewicz-Górka et al. 2015). Aluminum occurs commonly in many areas of our daily life, such as food additives, vaccine adjuvants, drugs components, cosmetics and cooking equipment, and might be constituents or contaminants in many food products as dairy, wine, juice, tea, sea foods and drinking water (López et al. 2002; Schiavo et al. 2008; Arnich et al. 2012). Absorption of aluminum into body by skin or by the respiratory or digestive system especially impairs the human central nervous system (Bragança et al. 2012; Niu 2018). Notable damage is seen in cognitive impairment in Al-exposed people, Alzheimer's disease and other neurodegenerative disorders (Flaten 1990; Ferreira et al. 2008; Li et al. 2017). Iron is one of the main elements used by man for centuries in many fields of human activity including mining, industry, water supply network (contamination that due to corrosion of pipes) and wastewater treatment systems (Kim et al. 2019b). Excessive Fe^{2+} increases oxidative stress via production of ROS, and might cause neurodegenerative diseases, including Alzheimer's disease (Xu et al. 2012; Pirpamer et al. 2016; Huat et al. 2019).

REMOVAL OF THE HEAVY METALS IONS FROM WASTEWATERS

One of the most common pathways for ingestion of heavy metals is through drinking water causing potentially harmful effects on human health. The presence of heavy metals in drinking water above recommended limits as laid down by the regulatory authorities in different countries of the world (Table 1), cause serious adverse effect on human health and is growing as a significant health concern for the authorities and healthcare professionals (Tchounwou et al. 2012). Supply of safe drinking water is crucial to human life, and safe drinking water should not impose a significant risk to humans (Chowdhury et al. 2016). Industrial processes produce wastes that are mostly discharged into the environment (Lichtfouse et al. 2012). Industrial activities, especially electroplating, metal smelting and chemical industries, and manufacturing processes are few sources of anthropogenic heavy metals in water (Jaishankar et al. 2014). Heavy metals are often present at high concentration in industrial wastewater and due to their ability to accumulate in aquatic organisms, then being easily transferred through the food chain to humans (Dang et al. 2009). There are many conventional methods for the purification of wastewater based on physico-chemical removal processes such as ultrafiltration, chemical precipitation, membrane filtration, electroflotation, electrodialysis, ion exchange, reverse osmosis and adsorbents (Cimino et al. 2005; Akbal and Camcı 2011; Shaheen et al. 2013; Ho et al. 2017). Due to the many disadvantages and relatively expensive operational procedures, some new technologies have invented and applied (Barakat 2011; Bhatnagar et al. 2015). This new technology uses agricultural waste materials as bio-adsorbents of heavy metals as a low cost and highly efficient technology. Because of the presence of many functional groups, this plant material can be used as good adsorbent for removal of cations and anions from aqueous solutions (De Gisi et al. 2016). For example, the following nonconventional adsorbents have been used: portulaca plant biomass (Dubey et al. 2014), apple waste (Marañón and Sastre 1991; Enniya et al. 2018), peanut hull carbon (Periasamy and Namasivayam 1995), doum-palm seed coat (El-Sadaawy and Abdelwahab 2014), and red mud (Apak et al. 1998).

Table 1. Permissible limits of the most dangerous heavy metals in drinking water establish by different authorities around the world

Metal concentration [µg/l]	WHO	EU	US-EPA	Classification group evaluated by IARC
Arsenic	3	10	10	Group 1
Cadmium	5	5	5	Group 1
Lead	10	10	15	Group 2B
Nickel	70	20	-	Group 1
Mercury	6	1	2	-
References	(World Health Organization 2011)	EU Council Directive	(US EPA 2015)	(International Agency for Research on Cancer 2012)

No matter which method is used to purify water, another task is choosing the proper analytical method that allows determination of the concentration of crucial heavy metals for human health in drinking water and their accordance with regulations of different authorities (Table 1). Moreover, analysis of sea, surface or ground waters allows researchers to examine the condition of the natural environment and degree of pollution. Water, especially drinking water, may by analyzed in different ways including color, turbidity, odor, pH, taste, hardness, conductivity, oxidation–reduction potential and finally presence of inorganic and organic constitutes (Gunnarsdottir et al. 2016; Udhayakumar et al. 2016; Heibati et al. 2017).

ANALYTICAL METHODS FOR QUANTIFICATION AND IDENTIFICATION OF HEAVY METALS

Nowadays, diverse detection methods have been used to quantify the concentration of heavy metal ions, including colorimetric methods (Hung et al. 2010; Chen et al. 2015; Terra et al. 2017; Kanellis 2018), electrochemical determination like anodic stripping voltammetry, (Huang and He 2013; Gherasim et al. 2014; Xing et al. 2016; Deshmukh et al. 2017; Lv et al. 2017) and spectrometric techniques such as: atomic absorption spectrometry with different types of atomizers including: flame (Siraj and Kitte 2013), electrothermal (graphite tube) (Liang et al. 2016; Han et al. 2017; Liu et al. 2018b), hydride, and cold vapor techniques (Subramanian 1989; Liu et al. 2005), flame photometry; inductively coupled plasma emission spectrometry and inductively coupled plasma mass spectrometry (Xu et al. 2007; Liu et al. 2015), and inductively coupled plasma-optical emission *spectrometry* (Sereshti et al. 2012) and fluorescence spectrometry (Liu et al. 2012).

ATOMIC ABSORPTION SPECTROMETRY

Atomic absorption spectrometry (AAS) is one of the most popular, reliable and suitable methods of heavy metal analysis in environmental samples including water (Table 2). It's a spectroanalytical procedure in which free metallic ions in the gaseous stage absorb optical radiation (light) at a specific wavelength to produce a measurable signal, which allows quantitative determination of analyzing elements (Fernández et al. 2018).

Table 2. Analytical methods use to determination of different heavy metals in water

Heavy metals	Analytical Method	References
As	Flame atomic absorption spectrophotometry	(Baig et al. 2012)
Cu, Zn, Pb	Flame atomic absorption spectrophotometry	(Siraj and Kitte 2013)
Pb	Flame atomic absorption spectrophotometry	(Zietz et al. 2007)
Cu, Co, Cd, Pb and Cr	Flame atomic absorption spectrophotometry	(Soylak et al. 2007)
Cr, Ni, Cu	Flame atomic absorption spectrophotometry	(Dong et al. 2015)
Mn, Cu, Co, Ni, Cd, Pb	Flame atomic absorption spectrophotometry	(Duran et al. 2007)
Hg	Cryogenic Gas Chromatography with Cold Vapour Atomic Fluorescence Detection	(Bloom and Fitzgerald 1988; Bloom 1989)
Methylmercury	Isothermal gas chromatography, thermal decomposition, and cold-vapor atomic fluorescence detection	(Gagnon et al. 1996; Desrosiers et al. 2006)
Fe	Flame atomic absorption spectrophotometer (FAAS)	(Brahman et al. 2013)
Zn, Fe, As, Cu, Mg, Cd	Fame atomic absorption spectrophotometry	(Kumar et al. 2017)
Fe, Pb, Zn, Cu, Ni, Cd, Cr	Atomic absorption spectrophotometry	(Amrane and Bouhidel 2019)
As, Hg, Zn, Fe, Cu, Cr, Mg, Cd	Atomic absorption spectrophotometry	(Kumar et al. 2016)
As	Electrothermal atomic absorption spectrometry	(Haghnazari et al. 2018)
As, Pb, Hg	Electrothermal atomic absorption spectrometry	(Mafa et al. 2016)
Cd, Cu, Pb, As	Electrothermal atomic absorption spectrometry	(Shirani et al. 2019)
As	Electrothermal atomic absorption spectrometry	(Hassanpoor et al. 2015)
Al, Cd, Co, Cu, Fe, Fe, Ni, Pb	Inductively coupled plasma optical emission spectrometry	(Losev et al. 2015)
Cu, Fe, Zn, Cd, Co, Mn, Ni	Inductively coupled plasma mass spectrometry	(Sánchez Trujillo et al. 2012)
Pt, Pd, Rh	Inductively coupled plasma mass spectrometry	(Liu et al. 2018a)

Heavy metals	Analytical Method	References
As, Sb, Hg, Bi	Atomic fluorescence spectroscopy	(Zhang et al. 2015)
Cd, Cr, Cu, Ni, Pb, Zn	Laser-induced breakdown spectroscopy	(Zhao et al. 2019)
Sb	Hydride generation atomic absorption spectrometry	(Erdem and Eroğlu 2005)
Cu, Cd, Pb	Bioelectronic tongues- Array of four gold electrodes modified with three peptides. Single gold electrode modified with the three peptides	(Chow et al. 2006)
Pb, Cd, Zn	Bioelectronic tongues- Array of three peptide-modified electrodes	(Serrano et al. 2014)
Cu	Electrochemical determination- mesoporous carbon electrode	(Huang and He 2013)
Hg, Pb	Electrochemical determination- carboimidazole grafted reduced graphene oxide) modified electrode	(Xing et al. 2016)
Pb	Electrochemical determination- microfabricated silver working electrode	(Jung et al. 2011)
Hg	Electrochemical determination- Gold nanoparticles decorated carbon fiber mat	(Li et al. 2014)
Hg, Pb, Cd	Electrochemical determination- Terephthalic acid capped iron oxide nanoparticles	(Deshmukh et al. 2017)
Cu, Cr, Co	Colorimetric detection	(Jeong and Kim 2015)
Hg	Colorimetric detection	(Kim et al. 2019a)
Hg, Ag, Pb	Colorimetric detection	(Hung et al. 2010)
Se	Colorimetric detection	(Mathew and Narayana 2006)

Table 2. (Continued)

Heavy metals	Analytical Method	References
As, Cr	High performance liquid chromatography inductively coupled plasma mass spectrometry	(Marcinkowska et al. 2015)
Cr	High performance liquid chromatography hyphenated to inductively coupled plasma mass spectrometry	(Markiewicz et al. 2015)
As, Se	High performance liquid chromatography (HPLC) hyphenated to inductively coupled plasma mass spectrometry (ICP-DRC-MS)	(Guerin et al. 1997)
Cd	Flame atomic absorption spectrometry	(Ferreira et al. 2007)
Cd	Graphite furnace atomic absorption spectrometry	
Mn, Co, Ni, Cu, Cr, Fe, Al, Zn, Sn, As, Cd, Pb, Sb, Hg	Inductively coupled plasma mass spectrometry	(Ghuniem et al. 2018)
Cr, Fe, Zn, As	Inductively coupled plasma mass spectrometry	(Dwivedi et al. 2019)
Ce, Dy, La, Sm, Y, U	Inductively coupled plasma optical emission spectrometry	(Bahramifar and Yamini 2005)
Bi, Cd, Co, Cu, Fe, Ga, In, Ni, Pb, Tl, Zn	Inductively coupled plasma-optical emission spectrometry	(Sereshti et al. 2012)
As	Hydride generation atomic absorption spectrometry	(Uluozlu et al. 2010)

Table 3. Guidelines of WHO related to proper analytical method to heavy metals determination in drinking water (World Health Organization 2011)

Metal	Limit of detection and methods	Provisional guideline value	IARC group
As	µg/l by ICP-MS; 2 µg/l by hydride generation AAS or flame AAS	10 µg/l	1
Cd	0.01 µg/l by ICP-MS; 2 µg/l by flame AAS	3 µg/l	1
Cr	0.05–0.2 µg/l for total chromium by AAS	50 µg/l	Cr (III)- 3, Cr (VI)- 1
Pb	1 µg/l by AAS; practical quantification limit in the region of 1–10 µg/	10	2B
Hg	0.05 µg/l by cold vapour AAS; 0.6 µg/l by ICP; 5 µg/l by flame AAS	6	-
Ni	0.1 µg/l by ICP-MS; 0.5 µg/l by flame AAS; 10 µg/l by ICP-AES	70	1
Cu	0.02–0.1 µg/l by ICP-MS; 0.3 µg/l by ICP–optical emission spectroscopy; 0.5 µg/l by flame AAS	2000	-

In order to analyze a sample by AAS, the analyte has to be first converted into gaseous atoms with an atomizer. Currently, the most commonly used atomizers are flame (spectroscopic) and electrothermal (graphite tubes). Atomization involves introducing a solution of a sample into a flame, while electrothermal atomization is where a drop of sample is placed into a graphite tube that is then heated electrically. Other atomizers, which have been recently introduced such as the hydride generation atomizer and cold vapor atomization, might also be useful for special purposes (Duran et al. 2007).

FLAME ATOMIC ABSORPTION SPECTROMETRY

In flame atomic absorption spectrometry (FAAS), metal ions in solution are converted to an atomic state by means of a flame. The technique is based on the principle that ground state metals absorb light at specific wavelengths. The absorbance is proportional to the number (concentration) of free, ground state atoms in a sample according to linearity of a standard curve. Due to the fact that each metal has a characteristic absorption wavelength, FAAS allows simultaneous analysis of a complex sample that contains many elements without the need for a separation process (Fernández et al. 2018).

By using FAAS, the following metals may be assay: antimony, bismuth, cadmium, calcium, cesium, chromium, cobalt, copper, gold, iridium, iron, lead, lithium, magnesium, manganese, nickel, palladium, platinum, potassium, rhodium, ruthenium, silver, sodium, strontium, thallium, tin, and zinc (Cassella et al. 1999; Narin and Soylak 2003; Baig et al. 2012; Brahman et al. 2013; Siraj and Kitte 2013; Dong et al. 2015; Zhang et al. 2015).

The application of FAAS to a direct determination of heavy metal ions at trace levels is restrained due to their low concentrations and matrix interferences. To eliminate this restriction, separation and pre-concentration methods are used (Section 4.1.4.).

ELECTROTHERMAL ATOMIC ABSORPTION SPECTROMETRY

Electrothermal atomic absorption (ETAAS) or graphite furnace atomic absorption spectroscopy (GFAAS) is a sensitive technique that offers detection limits which are about a factor of 20–1000 lower than for FAAS without extraction or sample concentration. It is a suitable standard method for determination of many trace elements, especially for background values, and for unpolluted samples, such as fresh water and biological materials. In ETAAS, samples are deposited in a small graphite or pyrolytic carbon coated graphite tube which can then be heated to vaporize and atomize the analyte. Compared to FAAS, atomizer modification increases sensitivity results from a higher atom density within the furnace. GFAAS provides detection limits as low as 1.0 µg/L for most elements. An additional advantage of GFAAS is that it requires a very small sample volume and problems of interference are minimized. This extremely high sensitivity technique is extremely prone to contamination and special care in sample handling and analysis is necessary (Jalbani et al. 2006; Haghnazari et al. 2018).

ETAAS allows to determine very low quantities of aluminum, antimony, arsenic, barium, beryllium, cadmium, chromium, cobalt, copper, iron, lead, manganese, molybdenum, nickel, selenium, silver, and tin (Elçi et al. 2008; Asadollahzadeh et al. 2014; Zounr et al. 2018; Shirani et al. 2019).

INDUCTIVELY COUPLED PLASMA-MASS SPECTROMETRY

Inductively coupled plasma-mass spectrometry (ICP-MS) is a powerful tool for detecting a large range of trace concentrations of elements such as metals in food, water, and environmental samples. ICP was introduced in 1983 in many laboratories and is based on atomic emission spectrometry (Batsala et al. 2012). Compared with AAS and ETAAS, ICP-MS offers several advantages like as greater sensitivity, speed, and precision. ICP-MS is a plasma source in which the energy is supplied by electric currents produced by electromagnetic induction. The ICP generates a high-temperature plasma at 10,000 degrees Celsius, through which the sample is passed. During this passage, at such high temperature, the components of the sample are ionized and directed into the MS. Then, a typical mass spectrometry analysis is performed where ions are sorted and separated according to their mass-to-charge ratio, and the detector identifies and quantifies each ion proportional to the quantity. Plasma is more beneficial than some methods such as flame ionization because it allows ionization to occur in a chemically inert environment, preventing oxide formation, and the ionization is more complete. One of the major advantages of ICP-MS is extremely low detection limits for a wide variety of elements. Heavy metals can be measured at part per trillion levels (Thomas 2004; Tanase et al. 2014; Ghuniem et al. 2018). ICP-MS is applicable for the detection of metals in different samples including drinking water, wastewater, soil science, food science, metallurgy.

This method is suitable for aluminum, antimony, arsenic, barium, beryllium, cadmium, chromium, cobalt, copper, lead, manganese, molybdenum, nickel, selenium, silver, strontium, thallium, uranium, vanadium, and zinc and many other elemental analytes (See Table 2).

Other spectrometric methods have lower usage and are limited to determination of Hg, Cd by cold-vapor atomic absorption spectroscopy and As, Cd, Bi, Ge, Sb, Se, Sn, Te and even In, Pb and Tl by hydride generation atomic absorption spectrometry (HGAAS) (Bakırdere et al. 2016; Büyükpınar et al. 2018). Hydride generation atomic absorption spectrometry is applicable to the determination of these ions by conversion to their hydrides by sodium borohydride reagent (Matusiewicz 2003).

LIMITATIONS OF ATOMIC ABSORPTION SPECTROSCOPY

Spectrometric methods are expensive, painstaking, time consuming and unsuitable for direct analysis of sample because of the complex operating procedure and advanced equipment. The other disadvantages are that only solutions can be analyzed and many lamps are needed (one lamp for one periodic element). In addition, it is difficult to determine elements occurring in high concentration and to do so requires relatively large sample quantities (1-2 mL). On the other hand, the method is highly precise, inexpensive, relatively easy to use and many elements may be assayed (Ferreira et al. 2007).

Colorless, single phase, no odor and transparent samples (mostly drinking water) can be analyzed directly by AAS (flame or electrothermal) or by ICP-MS for total metals without digestion. Digestion is a procedure that allows for a reduction in interference by organic matter and conversion of metals associated with particulates to a form that can be determined by spectroscopy techniques. There are two main digestion methods: the acid digestion technique (nitric, perchloric, hydrochloric, hydrofluoric, or sulfuric acid) and the microwave digestion procedure. The microwave method is a closed-vessel procedure and thus is expected to provide improved precision.

Samples containing organic material or particulates often require more advanced pretreatment procedures before spectroscopic analysis. The most suitable sample preparation techniques for further AAS method requirements are: liquid–liquid extraction (Okamoto et al. 2000; Wang et al. 2011), ion-exchange (Jiang et al. 2005), co-precipitation (Liu et al. 2000; Saracoglu et al. 2006; Divrikli et al. 2008; Bahadır et al. 2014), solid-phase extraction (Álvarez et al. 2007; Rajesh et al. 2007; Duran et al. 2007; Daşbaşı et al. 2015), flotation (Polat and Erdogan 2007), electrochemical deposition (Matusiewicz and Lesiński 2002; Tonini and Ruotolo 2017), and cloud point extraction (Zhao et al. 2012; Peng et al. 2015). Solid-phase extraction (SPE) is a cost-effective and reliable technique which overcomes many drawbacks of the traditional liquid–liquid extraction. SPE is very useful in the pre-concentration procedure of heavy metals, especially at low concentration and low sample volume. It generates smaller amounts of organic solvent waste and provides a more stable chemical form of the target elements on the solid surface (Duran et al. 2007; Yilmaz et al. 2014).

In recent years, various sample preparation methods, based on solvent microextraction approaches such as single-drop microextraction (Cheng et al. 2013; Rutkowska et al. 2014), hollow fiber-based liquid-phase microextraction (Jiang et al. 2009) dispersive liquid–liquid microextraction ((Moghimi 2008; Lai et al. 2016), countercurrent liquid–liquid microextraction (Mitani and Anthemidis 2015) and dispersive liquid–liquid microextraction based on solidification of a floating organic drop (Ghambarian et al. 2010) have been developed for the extraction and pre-concentration of heavy metals in different samples type including water. This sample preparation methods are next combined with different types of AAS and chromatography techniques (Losev et al. 2015; Gong et al. 2016)

ELECTROCHEMICAL TECHNIQUES

Electrochemical methods are attractive because of their great sensitivity and low cost. Electrochemical techniques measure different electrical signals. Amperometric, potentiometric, voltammetric, impedance measurement, coulometric and electro chemiluminescent techniques can be distinguished (Cui et al. 2015). Among them, voltammetry techniques are most commonly used in the field of heavy metal ion analysis in various complex environmental samples (Lu et al. 2018).

Stripping voltammetry (SV) constitutes a subdivision of voltammetry and has been demonstrated to be a very rapid and sensitive electro-analytical technique. For some metal ions it is even 10 to 100 times more sensitive than ETAAS, enabling detection limits in ppb and even the ppt range. Stripping analysis involves pre-concentration of a metal phase onto a solid electrode surface or into Hg (liquid) at negative potentials and selective oxidation of each metal phase species during an anodic potential sweep. Anodic stripping voltammetry (ASV) is one of the most sensitive electrochemical method used for the detection of heavy metal ions (Achterberg and Braungardt 1999; Achterberg et al. 2018; Abollino et al. 2019). Stripping voltammetry doesn't require a sample preparation step, utilizes cheap instrumentation and allows simultaneous examination of four to six trace metals. On the other hand, ASV has some disadvantages related to longer analysis times compere to spectroscopic methods, a limitation of application to amalgam-forming metals and risk of interferences (Achterberg and Braungardt 1999; Wang et al. 2000; Lu et al. 2018).

Anodic stripping voltammetry includes two steps: pre-concentration and dissolution (Pujol et al. 2014). First pre-concentration of an analyte at the working electrode surface using adsorption or Faraday's reaction, in order to lower the detection limit for that analyte in the sample solution, and then obtaining a correlated signal by anodic stripping voltammetry. Next, the dissolution step is performed by sweeping the electrode potential in the anodic direction (Gao and Huang 2013). At present, ASV is one of the most popular techniques used for online monitoring of heavy metals in water and in industrial wastewater (Hung et al. 2010; Zinoubi et al. 2017).

Earlier, working electrodes predominantly were constructed using mercury as an interface material. However, mercury electrodes are mechanically unstable and toxic, and because of this electrodes are unsuitable for automated analysis of heavy metals ions (Dell'Aglio et al. 2011; Unnikrishnan et al. 2013). A lot of research projects are focused on building chemically modified electrodes based on various bare solid electrodes of gold, silver, carbon, platinum and iridium modified using different interface materials like as electrochemical biosensors, polymers, metal film (bismuth film electrodes), metal oxides, nanomaterials and carbon nanotubes (Wang et al. 2000; Arduini et al. 2010; Bansod et al. 2017). Research and development of biosensors is becoming the most extensively studied discipline because they are easy, rapid, low-cost, highly sensitive, and highly selective. Electronic tongues are multisensor systems that consist of a number of low-selective sensors and use advanced mathematical procedures for signal processing based on pattern recognition and/or multivariate analysis (principal component analysis, artificial neural networks) (Vlasov et al. 2005). This approach allows simultaneous determination of many analytes in a complex sample, without any pretreatment step (Cetó et al. 2016). Electrochemical techniques are suitable for analysis of Ag, As, Cd, Cr, Cu, Hg, Pb, Zn and many more ions in liquid samples (Achterberg et al. 2018).

COLORIMETRIC METHODS

Colorimetry is usually applicable to specific heavy metal ions determination in water, which is really difficult in concurrent multiple element detection and where interferences are known not to compromise method accuracy (Chen et al. 2015; Terra et al. 2017). Colorimetry is a scientific technique that is used to determine the concentration of colored compounds in solutions by the application of the Beer–Lambert law, which states that the concentration of a solute is proportional to the absorbance. Colorimetry requires a colorimeter- a device used to measure absorbance of a solution a specific wavelength of light (Pearce 2014). Colorimetric methods for the determination of heavy metal ions demand specific chromogenic reagents that create colored compounds with them in solution (Jeong and Kim 2015). Colorimetric detection is really attractive because the outcomes can be easily interpreted with the naked eye, it is user-friendly, and is either equipment-free or uses low-cost instruments (Xiong et al. 2018).

Se, Cu, Cr, Co, Hg, Ag, Pb, Al are the examples of ions possible to determine by the colorimetric technique (Table 2).

CHROMATOGRAPHIC TECHNIQUES

High performance liquid chromatography (HPLC) hyphenated to inductively coupled plasma mass spectrometry (ICP-MS) has been recognized recently as the most effective and widely employed instrumental technique for speciation analysis including heavy metals analysis. ICP-MS is well matched as a detector for analyzing organometallic compounds as well as inorganic that can be separated by

HPLC, especially when an aqueous mobile phase is used. Coupling of HPLC with ICP- MS needs a direct interface between them, and the necessity of obtaining a consistent flow between the mobile phase through the chromatographic column and through the nebulizer (Delafiori et al. 2016). As in other liquid chromatography techniques, the type of column packing and the composition of the solid phase mainly depends on the nature of the sample such as polarity, solubility, and molecular weight. Manufacturers offer various types of columns, such as: reversed phase, reversed phase ion-pair, size exclusion, ion exchange, and chiral (Caruso and Montes-Bayon 2003). HPLC/ICP-MS doesn't require sample pretreatment process that reduces the probability of sample loss and allows to conduct multi-elemental analysis (Moreno et al. 2010; Sun et al. 2015).

ICP-DRC-MS allows for assay, frequently simultaneously, of following heavy metals ions: As, Se, Cr, Sb, Te, Mo, Pb, Hg, Cd, Te (Guerin et al. 1997; Roig-Navarro et al. 2001; Marcinkowska and Barałkiewicz 2016)

ANOTHER TECHNIQUES

There are many other analytical methods that are useful in the field of heavy metal analysis like: X-ray Fluorescence Spectrometry (O'Neil et al. 2015; Byers et al. 2019), neutron activation analysis (Eckhoff et al. 1973), fluorescence spectroscopy (Stedmon et al. 2011), Fourier-transform spectroscopy (Wang et al. 2017), inductively coupled plasma-atomic emission spectrometry (Sreenivasa Rao et al. 2002), laser-induced breakdown spectroscopy (Zhao et al. 2019), surface plasmon resonance spectroscopy (Wang et al. 2007).

IARC (International Agency for Research on Cancer) classification and permissible limits of Heavy Metals in Drinking water regulated by WHO (World Health Organization); US-EPA (U.S. Environmental Protection Agency) and EU (European Union).

Conclusion

The exposure of humans to heavy metals occurs by various routes including the consumption of contaminated food and water, inhalation of polluted air, occupational exposure at the workplace and skin contact. Some heavy metals such as Cd, Hg, Pb, Cr, As, Ni and many more, in high concentrations, can be harmful to the body causing many serious diseases such as cardiovascular disorders, diabetes, nervous system disorders, damage of kidneys, lungs or liver and even infertility.

It is critical to inform people about possible exposure and ways to implement appropriate educational programs concerning how to avoid further exposure. National and international cooperation is essential for developing proper strategies to avert the noxious impact of heavy metals. Especially important is that safe drinking water is provided to the community. Contaminated wastewater must be purified before release into the environment. Various methods are used for that purpose, such as adsorption on conventional and nonconventional adsorbents, membrane filtration, chemical precipitation, ion exchange and electrodialysis method. Many sophisticated analytical methods dedicated to an analysis of heavy metal ions even in trace concentration have been already invented and atomic absorption spectrometry techniques, chromatographic and electrochemical methods especially stripping voltammetry are still being developed. It is still necessary to develop rapid, very sensitive and selective techniques for heavy metals determination that may enable multicomponent simultaneous on-line analysis. Good responses to this demand seem to be biosensors, which affords precise chemical data in a timely and cost-efficient manner.

Acknowledgments

Dorota Bartusik-Aebisher acknowledges support from the National Center of Science NCN (New drug delivery systems-MRI study, Grant OPUS-13 number 2017/25/B/ST4/02481).

REFERENCES

Abollino O, Giacomino A, Malandrino M (2019) Voltammetry Stripping Voltammetry. In: Worsfold P, Poole C, Townshend A, Miró M (eds) *Encyclopedia of Analytical Science (Third Edition)*. Academic Press, Oxford, pp 238–257.

Achterberg EP, Braungardt C (1999) Stripping voltammetry for the determination of trace metal speciation and in-situ measurements of trace metal distributions in marine waters. *Anal Chim Acta* 400:381–397. doi: 10.1016/S0003-2670(99)00619-4.

Achterberg EP, Gledhill M, Zhu K (2018) Voltammetry—Cathodic Stripping. In: Reference Module in Chemistry, *Molecular Sciences and Chemical Engineering*. Elsevier.

Akbal F, Camcı S (2011) Copper, chromium and nickel removal from metal plating wastewater by electrocoagulation. *Desalination* 269:214–222. doi: 10.1016/j.desal.2010.11.001.

Álvarez P, Blanco C, Granda M (2007) The adsorption of chromium (VI) from industrial wastewater by acid and base-activated lignocellulosic residues. *J Hazard Mater* 144:400–405. doi: 10.1016/j.jhazmat.2006.10.052.

Amrane C, Bouhidel KE (2019) Analysis and speciation of heavy metals in the water, sediments, and drinking water plant sludge of a deep and sulfate-rich Algerian reservoir. *Environ Monit Assess* 191:73. doi: 10.1007/s10661-019-7222-9.

Apak R, Tütem E, Hügül M, Hizal J (1998) Heavy metal cation retention by unconventional sorbents (red muds and fly ashes). *Water Res* 32:430–440. doi: 10.1016/S0043-1354(97)00204-2.

Arduini F, Calvo JQ, Palleschi G, Moscone D, Amine A (2010) Bismuth-modified electrodes for lead detection. *TrAC Trends Anal Chem* 29:1295–1304. doi: 10.1016/j.trac.2010.08.003.

Arnich N, Sirot V, Rivière G, Jean J, Noël L, Guérin T, Leblanc J-C (2012) Dietary exposure to trace elements and health risk assessment in the 2nd French Total Diet Study. *Food Chem Toxicol Int J Publ Br Ind Biol Res Assoc* 50:2432–2449. doi: 10.1016/j.fct.2012.04.016.

Asadollahzadeh M, Tavakoli H, Torab-Mostaedi M, Hosseini G, Hemmati A (2014) Response surface methodology based on central composite design as a chemometric tool for optimization of dispersive-solidification liquid–liquid microextraction for speciation of inorganic arsenic in environmental water samples. *Talanta* 123:25–31. doi: 10.1016/j.talanta.2013.11.071.

Bahadır Z, Bulut VN, Ozdes D, Duran C, Bektas H, Soylak M (2014) Separation and preconcentration of lead, chromium and copper by using with the combination coprecipitation-flame atomic absorption spectrometric determination. *J Ind Eng Chem* 20:1030–1034. doi: 10.1016/j.jiec.2013.06.039.

Bahramifar N, Yamini Y (2005) On-line preconcentration of some rare earth elements in water samples using C18-cartridge modified with 1-(2-pyridylazo) 2-naphtol (PAN) prior to simultaneous determination by inductively coupled plasma optical emission spectrometry (ICP–OES). *Anal Chim Acta* 540:325–332. doi: 10.1016/j.aca.2005.03.058.

Baig JA, Kazi TG, Shah AQ, Kandhro GA, Afridi HI, Khan S, Kolachi NF, Wadhwa SK (2012) Arsenic speciation and other parameters of surface and ground water samples of Jamshoro, Pakistan. *Int J Environ Anal Chem* 92:28–42. doi: 10.1080/03067319.2010.496053.

Bakırdere S, Chormey DS, Büyükpınar Ç, San N, Keyf S (2016) Determination of Lead in Drinking and Wastewater by Hydride Generation Atomic Absorption Spectrometry. *Anal Lett* 49:1917–1925. doi: 10.1080/00032719.2015.1127380.

Bansod B, Kumar T, Thakur R, Rana S, Singh I (2017) A review on various electrochemical techniques for heavy metal ions detection with different sensing platforms. *Biosens Bioelectron* 94:443–455. doi: 10.1016/j.bios.2017.03.031.

Barakat MA (2011) New trends in removing heavy metals from industrial wastewater. *Arab J Chem* 4:361–377. doi: 10.1016/j.arabjc.2010.07.019.

Batsala M, Chandu B, Sakala B, Nama S, Domatoti S (2012) *Inductively Coupled Plasma Mass Spectrometry* (ICP-MS). 10.

Bernhoft RA (2012) Mercury toxicity and treatment: a review of the literature. *J Environ Public Health* 2012:460508. doi: 10.1155/2012/460508.

Beyersmann D, Hartwig A (2008) Carcinogenic metal compounds: recent insight into molecular and cellular mechanisms. *Arch Toxicol* 82:493. doi: 10.1007/s00204-008-0313-y.

Bhatnagar A, Sillanpää M, Witek-Krowiak A (2015) Agricultural waste peels as versatile biomass for water purification – A review. *Chem Eng J* 270:244–271. doi: 10.1016/j.cej.2015.01.135.

Bloom N (1989) Determination of Picogram Levels of Methylmercury by Aqueous Phase Ethylation, Followed by Cryogenic Gas Chromatography with Cold Vapour Atomic Fluorescence Detection. *Can J Fish Aquat Sci* 46:1131–1140. doi: 10.1139/f89-147.

Bloom N, Fitzgerald WF (1988) Determination of volatile mercury species at the picogram level by low-temperature gas chromatography with cold-vapour atomic fluorescence detection. *Anal Chim Acta* 208:151–161. doi: 10.1016/S0003-2670(00)80743-6.

Bragança VLC, Melnikov P, Zanoni LZ (2012) Trace Elements in Fruit Juices. *Biol Trace Elem Res* 146:256–261. doi: 10.1007/s12011-011-9247-y.

Brahman KD, Kazi TG, Afridi HI, Naseem S, Arain SS, Ullah N (2013) Evaluation of high levels of fluoride, arsenic species and other physicochemical parameters in underground water of two sub districts of Tharparkar, Pakistan: A multivariate study. *Water Res* 47:1005–1020. doi: 10.1016/j.watres.2012.10.042.

Büyükpınar Ç, Bekar B, Maltepe E, Chormey DS, Turak F, San N, Bakırdere S (2018) Development of a sensitive closed batch vessel hydride generation atomic absorption spectrometry method for the determination of cadmium in aqueous samples. *Instrum Sci Technol* 46:645–655. doi: 10.1080/10739149.2018.1439060.

Byers HL, McHenry LJ, Grundl TJ (2019) XRF techniques to quantify heavy metals in vegetables at low detection limits. *Food Chem X* 1:100001. doi: 10.1016/j.fochx.2018.100001.

Caruso JA, Montes-Bayon M (2003) Elemental speciation studies--new directions for trace metal analysis. *Ecotoxicol Environ Saf* 56:148–163.

Cassella RJ, Bitencourt DT, Branco AG, Ferreira SLC, Jesus DS de, Carvalho MS de, Santelli RE (1999) On-line preconcentration system for flame atomic absorption spectrometry using unloaded polyurethane foam: determination of zinc in waters and biological materials. *J Anal At Spectrom* 14:1749–1753. doi: 10.1039/A904997E.

Cetó X, Voelcker NH, Prieto-Simón B (2016) Bioelectronic tongues: New trends and applications in water and food analysis. *Biosens Bioelectron* 79:608–626. doi: 10.1016/j.bios.2015.12.075.

Chen G, Guo Z, Zeng G, Tang L (2015) Fluorescent and colorimetric sensors for environmental mercury detection. *The Analyst* 140:5400–5443. doi: 10.1039/c5an00389j.

Chen H, Zheng C, Tu C, Zhu YG (1999) Heavy metal pollution in soils in China: Status and countermeasures. *Ambio* 28:130–134.

Cheng K, Choi K, Kim J, Sung IH, Chung DS (2013) Sensitive arsenic analysis by carrier-mediated counter-transport single drop microextraction coupled with capillary electrophoresis. *Microchem J Complete*: 220–225. doi: 10.1016/j.microc.2012.07.005.

Chow E, Ebrahimi D, Gooding JJ, Hibbert DB (2006) Application of N-PLS calibration to the simultaneous determination of Cu^{2+}, Cd^{2+} and Pb2+ using peptide modified electrochemical sensors. *Analyst* 131:1051–1057. doi: 10.1039/B604690H.

Chowdhury S, Mazumder MAJ, Al-Attas O, Husain T (2016) Heavy metals in drinking water: Occurrences, implications, and future needs in developing countries. *Sci Total Environ* 569–570:476–488. doi: 10.1016/j.scitotenv.2016.06.166.

Chronopoulos J, Haidouti C, Chronopoulou-Sereli A, Massas I (1997) Variations in plant and soil lead and cadmium content in urban parks in Athens, Greece. *Sci Total Environ* 196:91–98. doi: 10.1016/S0048-9697(96)05415-0.

Cimino G, Cappello RM, Caristi C, Toscano G (2005) Characterization of carbons from olive cake by sorption of wastewater pollutants. *Chemosphere* 61:947–955. doi: 10.1016/j.chemosphere.2005.03.042.

Cui L, Wu J, Ju H (2015) Electrochemical sensing of heavy metal ions with inorganic, organic and bio-materials. *Biosens Bioelectron* 63:276–286. doi: 10.1016/j.bios.2014.07.052.

Council EEU. 1998. Council Directive 98/83/EC of 3 November 1998 on the Quality of Water Intended for Human Consumption.

Council Directive of 3 November 1998 on the Quality of Water Intended for Human Consumption, *Official Journal of European Communities*, L 330/32,5.12.98.

Dang VB, Doan HD, Dang-Vu T, Lohi A (2009) Equilibrium and kinetics of biosorption of cadmium(II) and copper(II) ions by wheat straw. *Bioresour Technol* 100:211–219. doi: 10.1016/j.biortech.2008.05.031.

Daşbaşı T, Saçmacı Ş, Ülgen A, Kartal Ş (2015) A solid phase extraction procedure for the determination of Cd(II) and Pb(II) ions in food and water samples by flame atomic absorption spectrometry. *Food Chem* 174:591–596. doi: 10.1016/j.foodchem.2014.11.049.

Delafiori J, Ring G, Furey A (2016) Clinical applications of HPLC-ICP-MS element speciation: A review. *Talanta* 153:306–331. doi: 10.1016/j.talanta.2016.02.035.

Dell'Aglio M, Gaudiuso R, Senesi GS, Giacomo AD, Zaccone C, Miano TM, Pascale OD (2011) Monitoring of Cr, Cu, Pb, V and Zn in polluted soils by laser induced breakdown spectroscopy (LIBS). *J Environ Monit* 13:1422–1426. doi: 10.1039/C0EM00780C.

Deshmukh S, Kandasamy G, Upadhyay RK, Bhattacharya G, Banerjee D, Maity D, Deshusses MA, Roy SS (2017) Terephthalic acid capped iron oxide nanoparticles for sensitive electrochemical detection of heavy metal ions in water. *J Electroanal Chem* 788:91–98. doi: 10.1016/j.jelechem.2017.01.064.

Desrosiers M, Planas D, Mucci A (2006) Short-term responses to watershed logging on biomass mercury and methylmercury accumulation by periphyton in boreal lakes. *Can J Fish Aquat Sci* 63:1734–1745. doi: 10.1139/f06-077.

Divrikli U, Soylak M, Elci L (2008) Determination of total chromium by flame atomic absorption spectrometry after coprecipitation by cerium

(IV) hydroxide. *Environ Monit Assess* 138:167–172. doi: 10.1007/s10661-007-9754-7.

Dong D, Liu X, Guo Z, Hua X, Su Y, Liang D (2015) Seasonal and Spatial Variations of Heavy Metal Pollution in Water and Sediments of China's Tiaozi River. *Pol J Environ Stud* 24:2371–2379. doi: 10.15244/pjoes/59276.

Dong J, Yang Q, Sun L, Zeng Q, Liu S, Pan J, Liu X (2011) Assessing the concentration and potential dietary risk of heavy metals in vegetables at a Pb/Zn mine site, China. *Environ Earth Sci* 64:1317–1321. doi: 10.1007/s12665-011-0992-1.

Dubey A, Mishra A, Singhal S (2014) Application of dried plant biomass as novel low-cost adsorbent for removal of cadmium from aqueous solution. *Int J Environ Sci Technol* 11:1043–1050. doi: 10.1007/s13762-013-0278-0.

Duran C, Gundogdu A, Bulut VN, Soylak M, Elci L, Sentürk HB, Tüfekci M (2007) Solid-phase extraction of Mn(II), Co(II), Ni(II), Cu(II), Cd(II) and Pb(II) ions from environmental samples by flame atomic absorption spectrometry (FAAS). *J Hazard Mater* 146:347–355. doi: 10.1016/j.jhazmat.2006.12.029.

Dwivedi R, Chelliah T, Khare D, Singh A (2019) Assessment of heavy metals in water of Ganga's riverine cities.

Eckhoff ND, Clack RW, Anthony HD, Gray AP (1973) Measurement of toxic heavy metals by neutron activation analysis. *Toxicol Appl Pharmacol* 24:178–181. doi: 10.1016/0041-008X(73)90137-3.

Elçi L, Divrikli Ü, Soylak M (2008) Inorganic arsenic speciation in various water samples with GFAAS using coprecipitation. *Int J Environ Anal Chem* 88:711–723. doi: 10.1080/03067310802094984.

El-Sadaawy M, Abdelwahab O (2014) Adsorptive removal of nickel from aqueous solutions by activated carbons from doum seed (Hyphaenethebaica) coat. *Alex Eng J* 53:399–408. doi: 10.1016/j.aej.2014.03.014.

Enniya I, Rghioui L, Jourani A (2018) Adsorption of hexavalent chromium in aqueous solution on activated carbon prepared from apple peels. *Sustain Chem Pharm* 7:9–16. doi: 10.1016/j.scp.2017.11.003.

Erdem A, Eroğlu AE (2005) Speciation and preconcentration of inorganic antimony in waters by Duolite GT-73 microcolumn and determination by segmented flow injection-hydride generation atomic absorption spectrometry (SFI-HGAAS). *Talanta* 68:86–92. doi: 10.1016/j.talanta.2005.04.041.

Fernández B, Lobo L, Pereiro R (2018) Atomic Absorption Spectrometry: Fundamentals, Instrumentation and Capabilities. In: Reference Module in Chemistry, Molecular Sciences and Chemical Engineering. *Elsevier,* p B9780124095472140000.

Ferreira PC, Piai K de A, Takayanagui AMM, Segura-Muñoz SI (2008) Aluminum as a risk factor for Alzheimer's disease. *Rev Lat Am Enfermagem* 16:151–157. doi: 10.1590/s0104-11692008000100023.

Ferreira SLC, de Andrade JB, Korn M das GA, Pereira M de G, Lemos VA, dos Santos WNL, Rodrigues F de M, Souza AS, Ferreira HS, da Silva EGP (2007) Review of procedures involving separation and preconcentration for the determination of cadmium using spectrometric techniques. *J Hazard Mater* 145:358–367. doi: 10.1016/j.jhazmat.2007.03.077.

Flaten TP (1990) Geographical associations between aluminium in drinking water and death rates with dementia (including Alzheimer's disease), Parkinson's disease and amyotrophic lateral sclerosis in Norway. *Environ Geochem Health* 12:152–167. doi: 10.1007/BF01734064.

Flora SJS, Mittal M, Mehta A (2008) Heavy metal induced oxidative stress & its possible reversal by chelation therapy. *Indian J Med Res* 128:501–523.

France) International Agency for Research on Cancer L, Working Group on the Evaluation of Carcinogenic Risks to Humans I, France L, Jun (1990) IARC. 1990. Chromium, nickel and welding. IARC Monographs on the Evaluation of the Carcinogenic Risks to Humans, Lyon, 49.

Gagnon C, Pelletier E, Mucci A, Fitzgerald WF (1996) Diagenetic behavior of methylmercury in organic-rich coastal sediments. *Limnol Oceanogr* 41:428–434. doi: 10.4319/lo.1996.41.3.0428.

Gao C, Huang X-J (2013) Voltammetric determination of mercury(II). TrAC Trends Anal Chem 51:1–12. doi: 10.1016/j.trac.2013.05.010.

Genestra M (2007) Oxyl radicals, redox-sensitive signalling cascades and antioxidants. *Cell Signal* 19:1807–1819. doi: 10.1016/j.cellsig. 2007.04.009.

Ghambarian M, Khalili-Zanjani MR, Yamini Y, Esrafili A, Yazdanfar N (2010) Preconcentration and speciation of arsenic in water specimens by the combination of solidification of floating drop microextraction and electrothermal atomic absorption spectrometry. *Talanta* 81:197–201. doi: 10.1016/j.talanta.2009.11.056.

Gherasim C-V, Křivčík J, Mikulášek P (2014) Investigation of batch electrodialysis process for removal of lead ions from aqueous solutions. *Chem Eng J* 256:324–334. doi: 10.1016/j.cej.2014.06.094.

Ghuniem MM, Korshed MA, Souaya ER (2018) Optimization and Validation of an Analytical Method for the Determination of Some Trace and Toxic Elements in Canned Fruit Juices Using Quadrupole Inductively Coupled Plasma Mass Spectrometer. *J AOAC Int.* doi: 10.5740/jaoacint.18-0022.

Gong T, Liu J, Liu X, Liu J, Xiang J, Wu Y (2016) A sensitive and selective sensing platform based on CdTe QDs in the presence of l-cysteine for detection of silver, mercury and copper ions in water and various drinks. *Food Chem* 213:306–312. doi: 10.1016/j.foodchem. 2016.06.091.

Guerin T, Astruc M, Batel A, Borsier M (1997) Multielemental speciation of As, Se, Sb and Te by HPLC-ICP-MS1Presented at the Symposium on Analytical Sciences (SAS) IV, Belgium, 3–5 June, 1996.1. *Talanta* 44:2201–2208. doi: 10.1016/S0039-9140(97)00036-2.

Gunnarsdottir MJ, Gardarsson SM, Jonsson GS, Bartram J (2016) Chemical quality and regulatory compliance of drinking water in Iceland. *Int J Hyg Environ Health* 219:724–733. doi: 10.1016/j.ijheh. 2016.09.011.

Guo H, Cui H, Fang J, Zuo Z, Deng J, Wang X, Zhao L, Wu B, Chen K, Deng J (2016) Nickel chloride-induced apoptosis via mitochondria-

and Fas-mediated caspase-dependent pathways in broiler chickens. *Oncotarget* 7: doi: 10.18632/oncotarget.12946.

Haghnazari L, Mirzaei N, Arfaeinia H, Karimyan K, Sharafi H, Fattahi N (2018) Speciation of As(III)/As(V) and Total Inorganic Arsenic in Biological Fluids Using New Mode of Liquid-Phase Microextraction and Electrothermal Atomic Absorption Spectrometry. *Biol Trace Elem Res* 183:173–181. doi: 10.1007/s12011-017-1118-8.

Han Q, Huo Y, Wu J, He Y, Yang X, Yang L (2017) Determination of Ultra-trace Rhodium in Water Samples by Graphite Furnace Atomic Absorption Spectrometry after Cloud Point Extraction Using 2-(5-Iodo-2-Pyridylazo)-5-Dimethylaminoaniline as a Chelating *Agent. Mol Basel Switz* 22: doi: 10.3390/molecules22040487.

Hassanpoor S, Khayatian G, Azar ARJ (2015) Ultra-trace determination of arsenic species in environmental waters, food and biological samples using a modified aluminum oxide nanoparticle sorbent and AAS detection after multivariate optimization. *Microchim Acta* 182:1957–1965. doi: 10.1007/s00604-015-1532-6.

Heibati M, Stedmon CA, Stenroth K, Rauch S, Toljander J, Säve-Söderbergh M, Murphy KR (2017) Assessment of drinking water quality at the tap using fluorescence spectroscopy. *Water Res* 125:1–10 . doi: 10.1016/j.watres.2017.08.020.

Ho S-H, Zhu S, Chang J-S (2017) Recent advances in nanoscale-metal assisted biochar derived from waste biomass used for heavy metals removal. *Bioresour Technol* 246:123–134. doi: 10.1016/j.biortech.2017.08.061.

Huang C-C, He J-C (2013) Electrosorptive removal of copper ions from wastewater by using ordered mesoporous carbon electrodes. *Chem Eng J* 221:469–475. doi: 10.1016/j.cej.2013.02.028.

Huat TJ, Camats-Perna J, Newcombe EA, Valmas N, Kitazawa M, Medeiros R (2019) Metal Toxicity Links to Alzheimer's Disease and Neuroinflammation. *J Mol Biol* 431:1843–1868. doi: 10.1016/j.jmb.2019.01.018.

Hung YL, Hsiung TM, Chen YY, Huang YF, Huang CC (2010) Colorimetric Detection of Heavy Metal Ions Using Label-Free Gold

Nanoparticles and Alkanethiols. *J Phys Chem* C 114:16329–16334. doi: 10.1021/jp1061573.

Hyder O, Chung M, Cosgrove D, Herman JM, Li Z, Firoozmand A, Gurakar A, Koteish A, Pawlik TM (2013) Cadmium exposure and liver disease among US adults. *J Gastrointest Surg Off J Soc Surg Aliment Tract* 17:1265–1273. doi: 10.1007/s11605-013-2210-9.

International Agency for Research on Cancer, Weltgesund heitsorganisation (eds) (2012) *IARC monographs on the evaluation of carcinogenic risks to humans, volume 100 C, arsenic, metals, fibres, and dusts: this publication represents the views and expert opinions of an IARC Working Group on the Evaluation of Carcinogenic Risks to Humans*, which met in Lyon, 17 - 24 March 2009. IARC, Lyon.

Jaishankar M, Tseten T, Anbalagan N, Mathew BB, Beeregowda KN (2014) Toxicity, mechanism and health effects of some heavy metals. *Interdiscip Toxicol* 7:60–72. doi: 10.2478/intox-2014-0009.

Jalbani N, Kazi TG, Arain BM, Jamali MK, Afridi HI, Sarfraz RA (2006) Application of factorial design in optimization of ultrasonic-assisted extraction of aluminum in juices and soft drinks. *Talanta* 70:307–314. doi: 10.1016/j.talanta.2006.02.045.

Järup L (2003) Hazards of heavy metal contamination. *Br Med Bull* 68:167–182. doi: 10.1093/bmb/ldg032.

Jeong U, Kim Y (2015) Colorimetric detection of heavy metal ions using aminosilane. *J Ind Eng Chem* 31:393–396. doi: 10.1016/j.jiec.2015.07.014.

Jiang H, Hu B, Chen B, Xia L (2009) Hollow fiber liquid phase microextraction combined with electrothermal atomic absorption spectrometry for the speciation of arsenic (III) and arsenic (V) in fresh waters and human hair extracts. *Anal Chim Acta* 634:15–21. doi: 10.1016/j.aca.2008.12.008.

Jiang Z-T, Yu JC, Liu H-Y (2005) Simultaneous Determination of Cobalt, Copper and Zinc by Energy Dispersive X-ray Fluorescence Spectrometry after Preconcentration on PAR-loaded Ion-Exchange Resin. *Anal Sci* 21:851–854. doi: 10.2116/analsci.21.851.

Jung W, Jang A, Bishop PL, Ahn CH (2011) A polymer lab chip sensor with microfabricated planar silver electrode for continuous and on-site heavy metal measurement. *Sens Actuators B Chem* 155:145–153. doi: 10.1016/j.snb.2010.11.039.

Kanellis VG (2018) Sensitivity limits of biosensors used for the detection of metals in drinking water. *Biophys Rev* 10:1415–1426. doi: 10.1007/s12551-018-0457-9.

Kim J, Oh JS, Park KC, Gupta G, Yeon Lee C (2019a) Colorimetric detection of heavy metal ions in water via metal-organic framework. *Inorganica Chim Acta* 486:69–73. doi: 10.1016/j.ica.2018.10.025.

Kim J-J, Kim Y-S, Kumar V (2019b) Heavy metal toxicity: An update of chelating therapeutic strategies. *J Trace Elem Med Bi*ol 54:226–231. doi: 10.1016/j.jtemb.2019.05.003.

Kumar A, Kaur M, Mehra R, Sharma DK, Mishra R (2017) Comparative Study of Radon Concentration with Two Techniques and Elemental Analysis in Drinking Water Samples of the Jammu District, Jammu and Kashmir, India. *Health Phys* 113:271–281. doi: 10.1097/HP.0000000000000644.

Kumar A, Kaur M, Sharma S, Mehra R, Sharma DK, Mishra R (2016) Radiation Dose Due to Radon and Heavy Metal Analysis in Drinking Water Samples of Jammu District, Jammu & Kashmir, India. *Radiat Prot Dosimetry* 171:217–222. doi: 10.1093/rpd/ncw062.

Kurniawan TA, Chan GYS, Lo W-H, Babel S (2006) Physico–chemical treatment techniques for wastewater laden with heavy metals. *Chem Eng J* 118:83–98. doi: 10.1016/j.cej.2006.01.015.

Lai G, Chen G, Chen T (2016) Speciation of As(III) and As(V) in fruit juices by dispersive liquid-liquid microextraction and hydride generation-atomic fluorescence spectrometry. *Food Chem* 190:158–163. doi: 10.1016/j.foodchem.2015.05.052.

Li D, Li J, Jia X, Wang E (2014) Gold nanoparticles decorated carbon fiber mat as a novel sensing platform for sensitive detection of Hg(II). *Electrochem Commun* 42:30–33. doi: 10.1016/j.elecom.2014.02.003.

Li F, Yuan Y, Meng P, Wu M, Li S, Chen B (2017) Probabilistic acute risk assessment of cumulative exposure to organophosphorus and carbamate pesticides from dietary vegetables and fruits in Shanghai populations. *Food Addit Contam Part A* 34:819–831. doi: 10.1080/19440049.2017.1279350.

Liang P, Kang C, Mo Y (2016) One-step displacement dispersive liquid-liquid microextraction coupled with graphite furnace atomic absorption spectrometry for the selective determination of methylmercury in environmental samples. *Talanta* 149:1–5. doi: 10.1016/j.talanta.2015.11.039.

Lichtfouse E, Schwarzbauer J, Robert D (eds) (2012) *Environmental chemistry for a sustainable world*. Springer, Dordrecht; New York.

Liu J, Chen H, Mao X, Jin X (2000) Determination of Trace Copper, Lead, Cadmium, and Iron in Environmental and Biological Samples by Flame Atomic Absorption Spectrometry Coupled to Flow Injection On-Line Coprecipitation Preconcentration Using DDTC-Nickel as Coprecipitate Carrier. *Int J Environ Anal Chem* 76:267–282. doi: 10.1080/03067310008034135.

Liu J-M, Lin L, Wang X-X, Lin S-Q, Cai W-L, Zhang L-H, Zheng Z-Y (2012) Highly selective and sensitive detection of Cu^{2+} with lysine enhancing bovine serum albumin modified-carbon dots fluorescent probe. *Analyst* 137:2637–2642. doi: 10.1039/C2AN35130G.

Liu K, Gao X, Li L, Chen C-TA, Xing Q (2018a) Determination of ultra-trace Pt, Pd and Rh in seawater using an off-line pre-concentration method and inductively coupled plasma mass spectrometry. *Chemosphere* 212:429–437. doi: 10.1016/j.chemosphere.2018.08.098.

Liu L, Zheng H, Xu B, Xiao L, Chigan Y, Zhangluo Y (2018b) In-situ pre-concentration through repeated sampling and pyrolysis for ultrasensitive determination of thallium in drinking water by electrothermal atomic absorption spectrometry. *Talanta* 179:86–91. doi: 10.1016/j.talanta.2017.10.003.

Liu Y, Chang X, Yang D, Guo Y, Meng S (2005) Highly selective determination of inorganic mercury(II) after preconcentration with Hg(II)-imprinted diazoaminobenzene–vinylpyridine copolymers. *Anal Chim Acta* 538:85–91. doi: 10.1016/j.aca.2005.02.017.

Liu Y, He M, Chen B, Hu B (2015) Simultaneous speciation of inorganic arsenic, selenium and tellurium in environmental water samples by dispersive liquid liquid microextraction combined with electrothermal vaporization inductively coupled plasma mass spectrometry. *Talanta* 142:213–220. doi: 10.1016/j.talanta.2015.04.050.

López FF, Cabrera C, Lorenzo ML, López MC (2002) Aluminium content of drinking waters, fruit juices and soft drinks: contribution to dietary intake. *Sci Total Environ* 292:205–213. doi: 10.1016/S0048-9697(01)01122-6.

Losev VN, Buyko OV, Trofimchuk AK, Zuy ON (2015) Silica sequentially modified with polyhexamethylene guanidine and Arsenazo I for preconcentration and ICP–OES determination of metals in natural waters. *Microchem J* 123:84–89. doi: 10.1016/j.microc.2015.05.022.

Lu Y, Liang X, Niyungeko C, Zhou J, Xu J, Tian G (2018) A review of the identification and detection of heavy metal ions in the environment by voltammetry. *Talanta* 178:324–338. doi: 10.1016/j.talanta.2017.08.033.

Lv ZL, Qi GM, Jiang TJ, Guo Z, Yu DY, Liu JH, Huang XJ (2017) A simplified electrochemical instrument equipped with automated flow-injection system and network communication technology for remote online monitoring of heavy metal ions. *J Electroanal Chem* 791:49–55. doi: 10.1016/j.jelechem.2017.03.012.

Mafa PJ, Idris AO, Mabuba N, Arotiba OA (2016) Electrochemical co-detection of As(III), Hg(II) and Pb(II) on a bismuth modified exfoliated graphite electrode. *Talanta* 153:99–106. doi: 10.1016/j.talanta.2016.03.003.

Marañón E, Sastre H (1991) Heavy metal removal in packed beds using apple wastes. *Bioresour Technol* 38:39–43. doi: 10.1016/0960-8524(91)90219-A.

Marcinkowska M, Barałkiewicz D (2016) Multielemental speciation analysis by advanced hyphenated technique – HPLC/ICP-MS: A review. *Talanta* 161:177–204. doi: 10.1016/j.talanta.2016.08.034.

Marcinkowska M, Komorowicz I, Barałkiewicz D (2015) Study on multielemental speciation analysis of Cr(VI), As(III) and As(V) in water by advanced hyphenated technique HPLC/ICP-DRC-MS. Fast and reliable procedures. *Talanta* 144:233–240. doi: 10.1016/j.talanta.2015.04.087.

Markiewicz B, Komorowicz I, Sajnóg A, Belter M, Barałkiewicz D (2015) Chromium and its speciation in water samples by HPLC/ICP-MS – technique establishing metrological traceability: A review since 2000. *Talanta* 132:814–828. doi: 10.1016/j.talanta.2014.10.002.

Markiewicz-Górka I, Januszewska L, Michalak A, Prokopowicz A, Januszewska E, Pawlas N, Pawlas K (2015) Effects of chronic exposure to lead, cadmium, and manganese mixtures on oxidative stress in rat liver and heart/Utjecaj kronične istodobne izloženosti olovu, kadmiju i manganu na oksidativni stres u jetri i srcu štakora. *Arch Ind Hyg Toxicol* 66:51–62. doi: 10.1515/aiht-2015-66-2515.

Mathew M, Narayana B (2006) An easy spectrophotometric determination of selenium using azure B as a chromogenic reagent. *Indian J Chem Technol* 4.

Matusiewicz H (2003) Chemical Vapor Generation with Slurry Sampling: A Review of Atomic Absorption Applications. *Appl Spectrosc Rev* 38:263–294. doi: 10.1081/ASR-120023948.

Matusiewicz H, Lesiński M (2002) Electrodeposition Sample Introduction for Ultra Trace Determinations of Platinum Group Elements (Pt, Pd, Rh, Ru) in Road Dust by Electrothermal Atomic Absorption Spectrometry. *Int J Environ Anal Chem* 82:207–223. doi: 10.1080/03067310290027785.

Mitani C, Anthemidis AN (2015) An automatic countercurrent liquid–liquid micro-extraction system coupled with atomic absorption spectrometry for metal determination. *Talanta* 133:77–81. doi: 10.1016/j.talanta.2014.04.091.

Moghimi A (2008) Preconcentration Ultra Trace of Cd(II) in Water Samples Using Dispersive Liquid-Liquid Microextraction with Salen (N,N'-Bis(Salicylidene)-Ethylenediamine) and Determination Graphite

Furnace Atomic Absorption Spectrometry. *J Chin Chem Soc* 55:369–376 . doi: 10.1002/jccs.200800054.

Moreno F, García-Barrera T, Gómez-Ariza JL (2010) Simultaneous analysis of mercury and selenium species including chiral forms of selenomethionine in human urine and serum by HPLC column-switching coupled to ICP-MS. *The Analyst* 135:2700–2705. doi: 10.1039/c0an00090f.

Narin I, Soylak M (2003) Enrichment and determinations of nickel(II), cadmium(II), copper(II), cobalt(II) and lead(II) ions in natural waters, table salts, tea and urine samples as pyrrolydine dithiocarbamate chelates by membrane filtration–flame atomic absorption spectrometry combination. *Anal Chim Acta* 493:205–212. doi: 10.1016/S0003-2670(03)00867-5.

Nies DH (1999) Microbial heavy-metal resistance. *Appl Microbiol Biotechnol* 51:730–750.

Niu Q (2018) Overview of the Relationship Between Aluminum Exposure and Health of Human Being. In: Niu Q (ed) Neurotoxicity of Aluminum. Springer Singapore, Singapore, pp 1–31.

Okamoto Y, Nomura Y, Nakamura H, Iwamaru K, Fujiwara T, Kumamaru T (2000) High preconcentration of ultra-trace metal ions by liquid–liquid extraction using water/oil/water emulsions as liquid surfactant membranes. *Microchem J* 65:341–346. doi: 10.1016/S0026-265X(00)00161-2.

O'Neil GD, Newton ME, Macpherson JV (2015) Direct identification and analysis of heavy metals in solution (Hg, Cu, Pb, Zn, Ni) by use of in situ electrochemical X-ray fluorescence. *Anal Chem* 87:4933–4940. doi: 10.1021/acs.analchem.5b00597.

Orlowski C, Piotrowski JK (2003) Biological levels of cadmium and zinc in the small intestine of non-occupationally exposed human subjects. *Hum Exp Toxicol* 22:57–63. doi: 10.1191/0960327103ht326oa.

Park J-D, Zheng W (2012) Human Exposure and Health Effects of Inorganic and Elemental Mercury. *J Prev Med Pub Health* 45:344–352. doi: 10.3961/jpmph.2012.45.6.344.

Pearce JM (2014) Chapter 6 - Digital Designs and Scientific Hardware. In: Pearce JM (ed) Open-Source Lab. Elsevier, Boston, pp 165–252.

Peng G, He Q, Zhou G, Li Y, Su X, Liu M, Fan L (2015) Determination of heavy metals in water samples using dual-cloud point extraction coupled with inductively coupled plasma mass spectrometry. *Anal Methods* 7:6732–6739. doi: 10.1039/C5AY00801H.

Periasamy K, Namasivayam C (1995) Removal of nickel(II) from aqueous solution and nickel plating industry wastewater using an agricultural waste: Peanut hulls. *Waste Manag* 15:63–68. doi: 10.1016/0956-053X(94)00071-S.

Pirpamer L, Hofer E, Gesierich B, De Guio F, Freudenberger P, Seiler S, Duering M, Jouvent E, Duchesnay E, Dichgans M, Ropele S, Schmidt R (2016) Determinants of iron accumulation in the normal aging brain. *Neurobiol Aging* 43:149–155. doi: 10.1016/j.neurobiolaging.2016.04.002.

Polat H, Erdogan D (2007) Heavy metal removal from waste waters by ion flotation. *J Hazard Mater* 148:267–273. doi: 10.1016/j.jhazmat.2007.02.013.

Rahman Z, Singh VP (2019) The relative impact of toxic heavy metals (THMs) (arsenic (As), cadmium (Cd), chromium (Cr)(VI), mercury (Hg), and lead (Pb)) on the total environment: an overview. *Environ Monit Assess* 191:419. doi: 10.1007/s10661-019-7528-7.

Rajesh N, Deepthi B, Subramaniam A (2007) Solid phase extraction of chromium(VI) from aqueous solutions by adsorption of its ion-association complex with cetyltrimethylammoniumbromide on an alumina column. *J Hazard Mater* 144:464–469. doi: 10.1016/j.jhazmat.2006.10.059.

Rehman K, Fatima F, Waheed I, Akash MSH (2018) Prevalence of exposure of heavy metals and their impact on health consequences. *J Cell Biochem* 119:157–184. doi: 10.1002/jcb.26234.

Roig-Navarro AF, Martinez-Bravo Y, López FJ, Hernández F (2001) Simultaneous determination of arsenic species and chromium(VI) by high-performance liquid chromatography–inductively coupled plasma-mass spectrometry. *J Chromatogr A* 912:319–327. doi: 10.1016/ S0021-9673(01)00572-6.

Rusyniak DE, Arroyo A, Acciani J, Froberg B, Kao L, Furbee B (2010) Heavy metal poisoning: management of intoxication and antidotes. EXS 100:365–396.

Rutkowska M, Dubalska K, Konieczka P, Namieśnik J (2014) Microextraction techniques used in the procedures for determining organomercury and organotin compounds in environmental samples. *Mol Basel Switz* 19:7581–7609. doi: 10.3390/molecules19067581.

Sánchez Trujillo I, Vereda Alonso E, García de Torres A, Cano Pavón JM (2012) Development of a solid phase extraction method for the multielement determination of trace metals in natural waters including sea-water by FI-ICP-MS. *Microchem J* 101:87–94. doi: 10.1016/j.microc.2011.11.003.

Saracoglu S, Soylak M, Peker DSK, Elci L, dos Santos WNL, Lemos VA, Ferreira SLC (2006) A pre-concentration procedure using coprecipitation for determination of lead and iron in several samples using flame atomic absorption spectrometry. *Anal Chim Acta* 575:133–137. doi: 10.1016/j.aca.2006.05.055.

Schiavo D, Neira JY, Nóbrega JA (2008) Direct determination of Cd, Cu and Pb in wines and grape juices by thermospray flame furnace atomic absorption spectrometry. *Talanta* 76:1113–1118. doi: 10.1016/j.talanta.2008.05.010.

Sereshti H, Entezari Heravi Y, Samadi S (2012) Optimized ultrasound-assisted emulsification microextraction for simultaneous trace multielement determination of heavy metals in real water samples by ICP-OES. *Talanta* 97:235–241. doi: 10.1016/j.talanta.2012.04.024.

Serrano N, Prieto-Simón B, Cetó X, del Valle M (2014) Array of peptide-modified electrodes for the simultaneous determination of Pb(II), Cd(II) and Zn(II). *Talanta* 125:159–166. doi: 10.1016/j.talanta.2014.02.052.

Shaheen SM, Eissa FI, Ghanem KM, Gamal El-Din HM, Al Anany FS (2013) Heavy metals removal from aqueous solutions and wastewaters by using various byproducts. *J Environ Manage* 128:514–521. doi: 10.1016/j.jenvman.2013.05.061.

Shirani M, Habibollahi S, Akbari A (2019) Centrifuge-less deep eutectic solvent based magnetic nanofluid-linked air-agitated liquid-liquid microextraction coupled with electrothermal atomic absorption spectrometry for simultaneous determination of cadmium, lead, copper, and arsenic in food samples and non-alcoholic beverages. *Food Chem* 281:304–311. doi: 10.1016/j.foodchem.2018.12.110.

Singh N, Kumar D, Sahu AP (2007) Arsenic in the environment: effects on human health and possible prevention. *J Environ Biol* 28:359–365.

Siraj K, Kitte SA (2013) Analysis of Copper, Zinc and Lead using Atomic Absorption Spectrophotometer in ground water of Jimma town of Southwestern Ethiopia. *Int J Chem Anal Sci* 4:201–204. doi: 10.1016/j.ijcas.2013.07.006.

Soylak M, Divrikli U, Saracoglu S, Elci L (2007) Membrane filtration – atomic absorption spectrometry combination for copper, cobalt, cadmium, lead and chromium in environmental samples. *Environ Monit Assess* 127:169–176. doi: 10.1007/s10661-006-9271-0.

Sreenivasa Rao K, Balaji T, Prasada Rao T, Babu Y, Naidu GRK (2002) Determination of iron, cobalt, nickel, manganese, zinc, copper, cadmium and lead in human hair by inductively coupled plasma-atomic emission spectrometry. *Spectrochim Acta Part B At Spectrosc* 57:1333–1338. doi: 10.1016/S0584-8547(02)00045-9.

Stedmon CA, Seredyńska-Sobecka B, Boe-Hansen R, Le Tallec N, Waul CK, Arvin E (2011) A potential approach for monitoring drinking water quality from groundwater systems using organic matter fluorescence as an early warning for contamination events. *Water Res* 45:6030–6038. doi: 10.1016/j.watres.2011.08.066.

Subramanian KS (1989) Determination of lead in blood by graphite furnace atomic absorption spectrometry--a critique. *Sci Total Environ* 89:237–250.

Sun J, Yang Z, Lee H, Wang L (2015) Simultaneous speciation and determination of arsenic, chromium and cadmium in water samples by high performance liquid chromatography with inductively coupled plasma mass spectrometry. *Anal Methods* 7:2653–2658. doi: 10.1039/ C4AY02813A.

Tanase IG, Popa DE, Udriştioiu GE, Bunaciu AA, Aboul-Enein HY (2014) Validation and Quality Control of an ICP-MS Method for the Quantification and Discrimination of Trace Metals and Application in Paper Analysis: An Overview. *Crit Rev Anal Chem* 44:311–327. doi: 10.1080/10408347.2013.863141.

Tchounwou PB, Yedjou CG, Patlolla AK, Sutton DJ (2012) Heavy metal toxicity and the environment. *Exp Suppl* 2012 101:133–164. doi: 10.1007/978-3-7643-8340-4_6.

Terra IAA, Mercante LA, Andre RS, Correa DS (2017) Fluorescent and Colorimetric Electrospun Nanofibers for Heavy-Metal Sensing. *Biosensors* 7: doi: 10.3390/bios7040061.

Thomas R (2004) Practical guide to ICP-MS. M. Dekker, New York, NY.

Tonini GA, Ruotolo LAM (2017) Heavy metal removal from simulated wastewater using electrochemical technology: optimization of copper electrodeposition in a membraneless fluidized bed electrode. *Clean Technol Environ Policy* 19:403–415. doi: 10.1007/s10098-016-1226-8.

Torres F, das Graças M, Melo M, Tosti A (2009) Management of contact dermatitis due to nickel allergy: an update. *Clin Cosmet Investig Dermatol* 2:39–48.

Udhayakumar R, Manivannan P, Raghu K, Vaideki S (2016) Assessment of physico-chemical characteristics of water in Tamilnadu. *Ecotoxicol Environ Saf* 134:474–477. doi: 10.1016/j.ecoenv.2016.07.014.

Uluozlu OD, Tuzen M, Mendil D, Soylak M (2010) Determination of As(III) and As(V) species in some natural water and food samples by solid-phase extraction on Streptococcus pyogenes immobilized on Sepabeads SP 70 and hydride generation atomic absorption spectrometry. *Food Chem Toxicol* 48:1393–1398. doi: 10.1016/j.fct. 2010.03.007.

Unnikrishnan VK, Nayak R, Aithal K, Kartha VB, Santhosh C, Gupta GP, Suri BM (2013) Analysis of trace elements in complex matrices (soil) by Laser Induced Breakdown Spectroscopy (LIBS). *Anal Methods* 5:1294–1300. doi: 10.1039/C2AY26006A.

US EPA O (2015) National Primary Drinking Water Regulations. In: US EPA. https://www.epa.gov/ground-water-and-drinking-water/national-primary-drinking-water-regulations. Accessed 9 Jul 2019.

Vlasov Yu, Legin A, Rudnitskaya A, Di NC, D'Amico A (2005) Nonspecific sensor arrays ("electronic tongue") for chemical analysis of liquids (IUPAC Technical Report). *Pure Appl Chem* 77:1965–1983. doi: 10.1351/pac200577111965.

Wang C, Li W, Guo M, Ji J (2017) Ecological risk assessment on heavy metals in soils: Use of soil diffuse reflectance mid-infrared Fourier-transform spectroscopy. *Sci Rep* 7:40709. doi: 10.1038/srep40709.

Wang H, Ma J, Wu Q, Luo X, Chen Z, Kou L (2011) Circulating B lymphocytes producing autoantibodies to endothelial cells play a role in the pathogenesis of Takayasu arteritis. *J Vasc Surg* 53:174–180. doi: 10.1016/j.jvs.2010.06.173.

Wang J, Lu J, Hocevar SB, Farias PAM, Ogorevc B (2000) Bismuth-Coated Carbon Electrodes for Anodic Stripping Voltammetry. *Anal Chem* 72:3218–3222. doi: 10.1021/ac000108x.

Wang S, Forzani ES, Tao N (2007) Detection of Heavy Metal Ions in Water by High-Resolution Surface Plasmon Resonance Spectroscopy Combined with Anodic Stripping Voltammetry. *Anal Chem* 79:4427–4432. doi: 10.1021/ac0621773.

World Health Organization (ed) (2011) Guidelines for drinking-water quality, 4th ed. *World Health Organization*, Geneva.

Xing H, Xu J, Zhu X, Duan X, Lu L, Zuo Y, Zhang Y, Wang W (2016) A new electrochemical sensor based on carboimidazole grafted reduced graphene oxide for simultaneous detection of Hg^{2+} and Pb2+. *J Electroanal Chem C*: 250–255 . doi: 10.1016/j.jelechem.2016.10.043.

Xiong Y, Li F, Wang J, Huang A, Wu M, Zhang Z, Zhu D, Xie W, Duan Z, Su L (2018) Simple multimodal detection of selenium in water and vegetable samples by a catalytic chromogenic method. *Anal Methods* 10:2102–2107. doi: 10.1039/C8AY00265G.

Xu J, Jia Z, Knutson MD, Leeuwenburgh C (2012) Impaired Iron Status in Aging Research. *Int J Mol Sci* 13:2368–2386. doi: 10.3390/ijms13022368.

Xu Y, Zhou J, Wang G, Zhou J, Tao G (2007) Determination of trace amounts of lead, arsenic, nickel and cobalt in high-purity iron oxide pigment by inductively coupled plasma atomic emission spectrometry after iron matrix removal with extractant-contained resin. *Anal Chim Acta* 584:204–209. doi: 10.1016/j.aca.2006.11.014.

Yilmaz V, Arslan Z, Hazer O, Yilmaz H (2014) Selective solid phase extraction of copper using a new Cu(II)-imprinted polymer and determination by inductively coupled plasma optical emission spectroscopy (ICP-OES). *Microchem J* 114:65–72. doi: 10.1016/j.microc.2013.12.002.

Zambelli B, Ciurli S (2013) Nickel and Human Health. In: Sigel A, Sigel H, Sigel RKO (eds) Interrelations between Essential Metal Ions and Human Diseases. Springer Netherlands, Dordrecht, pp 321–357.

Zambelli B, Uversky VN, Ciurli S (2016) Nickel impact on human health: An intrinsic disorder perspective. *Biochim Biophys Acta BBA - Proteins Proteomics* 1864:1714–1731. doi: 10.1016/j.bbapap.2016.09.008.

Zhang PP, Lyu SS, Zhu XH, Chen XG, Wu DD, Ye Y (2015) Assessment of Arsenic Contamination in and around a Plateau Lake: Influences of Groundwater and Anthropogenic Pollution. *Pol J Environ Stud* 24:2715–2725. doi: 10.15244/pjoes/59296.

Zhang X, Yang L, Li Y, Li H, Wang W, Ye B (2012) Impacts of lead/zinc mining and smelting on the environment and human health in China. *Environ Monit Assess* 184:2261–2273. doi: 10.1007/s10661-011-2115-6.

Zhao L, Zhong S, Fang K, Qian Z, Chen J (2012) Determination of cadmium(II), cobalt(II), nickel(II), lead(II), zinc(II), and copper(II) in water samples using dual-cloud point extraction and inductively coupled plasma emission spectrometry. *J Hazard Mater* 239–240:206–212. doi: 10.1016/j.jhazmat.2012.08.066.

Zhao NJ, Meng DS, Jia Y, Ma MJ, Fang L, Liu JG, Liu WQ (2019) On-line quantitative analysis of heavy metals in water based on laser-induced breakdown spectroscopy. *Opt Express* 27:A495–A506. doi: 10.1364/OE.27.00A495.

Zietz DBP, Laß J, Suchenwirth R (2007) Assessment and management of tap water lead contamination in Lower Saxony, Germany. *Int J Environ Health Res* 17:407–418. doi: 10.1080/09603120701628719.

Zinoubi K, Majdoub H, Barhoumi H, Boufi S, Jaffrezic-Renault N (2017) Determination of trace heavy metal ions by anodic stripping voltammetry using nanofibrillated cellulose modified electrode. *J Electroanal Chem* 799:70–77. doi: 10.1016/j.jelechem.2017.05.039.

Zounr RA, Tuzen M, Deligonul N, Khuhawar MY (2018) A highly selective and sensitive ultrasonic assisted dispersive liquid phase microextraction based on deep eutectic solvent for determination of cadmium in food and water samples prior to electrothermal atomic absorption spectrometry. *Food Chem* 253:277–283. doi: 10.1016/j.foodchem.2018.01.167.

Żukowska J, Biziuk M (2008) Methodological Evaluation of Method for Dietary Heavy Metal Intake. *J Food Sci* 73:R21–R29. doi: 10.1111/j.1750-3841.2007.00648.x.

EDITOR CONTACT INFORMATION

Dorota Bartusik-Aebisher
University of Rzeszów
Faculty of Medicine
Rzeszów, Poland
Email: dbartusik-aebisher@ur.edu.pl

INDEX

A

absorption spectroscopy, 153, 154
acid, 2, 5, 18, 19, 24, 93, 95, 96, 97, 111, 123, 129, 135, 144, 149, 155, 161, 165
acidic, 46, 94, 106
activated carbon, 6, 39, 48, 99, 101, 103, 166, 167
active site, 56, 63, 93
adsorption, 2, 5, 9, 12, 20, 22, 29, 40, 47, 66, 72, 76, 83, 88, 90, 91, 92, 93, 94, 95, 96, 97, 98, 99, 100, 101, 102, 103, 104, 105, 106, 107, 108, 109, 110, 111, 157, 160, 161, 167, 176
age, 116, 136, 143
agriculture, 5, 114
algae, 88, 89, 91, 106, 107
amalgam, 13, 20, 21, 157
amino, 12, 126
anaerobic sludge, 12, 29, 50
antimony, 152, 153, 154, 167
antioxidant, 47, 118, 131
aqueous solutions, 23, 35, 51, 94, 100, 101, 103, 104, 106, 107, 109, 110, 127, 146, 166, 168, 176, 178
aqueous suspension, 41
arsenic, 23, 24, 28, 30, 36, 44, 48, 79, 96, 109, 110, 114, 130, 132, 139, 140, 142, 143, 153, 154, 162, 163, 164, 166, 168, 169, 170, 173, 176, 177, 178, 179, 181
arteritis, 180
asbestos, 6, 32, 34
Asian countries, 92, 104
assessment, 25, 36, 40, 107, 130, 134, 135
assimilation, 34, 93, 96
atmosphere, 126
atomic absorption spectrometry, vii, ix, 142, 147, 148, 149, 150, 152, 154, 160, 162, 163, 164, 165, 166, 167, 168, 169, 170, 172, 174, 175, 177, 178, 179, 182
atomic emission spectrometry, 153, 159, 178, 181
authorities, 145, 146
autoantibodies, 180

B

bacteria, 3, 4, 5, 8, 14, 15, 16, 18, 19, 20, 23, 25, 30, 34, 35, 39, 41, 47, 48, 49, 88, 89
bacterium, 18, 24, 46, 58
barium, 153, 154
barriers, 41, 44

base, 40, 88, 97, 110, 161
beryllium, 153, 154
bioaccumulation, 22, 28, 36, 38, 92
bioavailability, 28, 124, 125, 130
biodegradation, 17, 20, 29, 31, 36, 37, 40, 50, 117, 126
biofuel, 13, 37
biomass, 5, 13, 26, 30, 43, 44, 87, 88, 89, 90, 91, 92, 99, 106, 117, 120, 122, 123, 124, 127, 146, 163, 165, 166, 169
biomolecules, 56, 118
bioremediation, 3, 19, 22, 27, 34, 35, 41, 43, 44, 45, 89, 90, 91, 93, 94, 99
biosafety, 68
biosensors, 157, 160, 171
biosorbents, 16, 26, 34, 45, 89, 91, 100, 101, 102, 106, 110
bio-sorption, 3, 5, 12, 15, 18, 19, 21, 22, 23, 24, 25, 30, 31, 35, 37, 39, 42, 43, 45, 88, 91, 92, 100, 102, 103, 104, 105, 106, 107, 108, 109, 110, 165
biosynthesis, 44, 83, 118, 122
biotechnology, 20, 35, 60
biotic, 8, 48, 51, 135
bismuth, 152, 157, 173
black tea, 97, 110
blood, 54, 104, 129, 134, 138, 144, 178
bones, 54, 102
brain, 143, 144, 176
Brazil, 40, 129
breakdown, 18, 149, 159, 165, 182

C

Ca^{2+}, 69, 71, 85, 144
cadmium, 2, 12, 18, 20, 25, 29, 31, 33, 36, 41, 45, 49, 101, 102, 104, 105, 109, 114, 128, 129, 130, 131, 132, 133, 134, 137, 138, 140, 142, 143, 152, 153, 154, 163, 164, 165, 166, 167, 174, 175, 176, 178, 179, 181, 182

calcium, 15, 97, 152
cancer, 54, 58, 117, 143
cancer cells, 54, 58
capillary, 74, 164
carbohydrates, 54, 126
carbon, 3, 21, 33, 35, 37, 45, 50, 56, 64, 76, 79, 99, 110, 124, 146, 149, 153, 157, 169, 171, 172
carbon nanotubes, 56, 76, 110, 157
carboxyl, 90, 96
carcinogenesis, 116, 135
cardiovascular disorders, 143, 160
Cd, 2, 4, 5, 12, 19, 37, 39, 47, 48, 51, 88, 89, 90, 91, 92, 93, 94, 95, 96, 98, 99, 100, 103, 107, 114, 116, 120, 121, 122, 123, 125, 127, 131, 132, 133, 142, 148, 149, 150, 151, 154, 158, 159, 160, 165, 166, 175, 176, 177
cell death, 13, 69
cell line, 117, 132
cellulose, 34, 89, 90, 182
central nervous system, 89, 144
cesium, 25, 42, 152
challenges, 54, 104
chelates, 123, 175
chemical, 5, 6, 17, 19, 23, 25, 42, 50, 53, 54, 63, 68, 71, 90, 93, 94, 96, 97, 99, 113, 115, 122, 125, 141, 142, 145, 155, 160, 171, 179, 180
chemical properties, 6, 71, 142
China, 26, 41, 47, 130, 135, 139, 164, 166, 181
chitosan, 3, 18, 48, 69, 72, 82, 83, 89
Chitosan, 72, 82, 83
chromatography, 49, 148, 156, 159, 163
chromium, 15, 17, 19, 26, 29, 30, 32, 35, 38, 39, 40, 42, 43, 44, 46, 47, 89, 91, 92, 102, 103, 104, 106, 110, 114, 128, 130, 136, 142, 151, 152, 153, 154, 161, 162, 166, 167, 176, 177, 178, 179
coal, 18, 98, 101, 107

cobalt, 38, 65, 109, 152, 153, 154, 175, 178, 181
community, 36, 49, 65, 160
composites, 63, 72, 107
composition, 5, 36, 37, 49, 54, 91, 96, 159
compost, 3, 89, 92, 111, 125
compounds, 3, 13, 17, 22, 32, 54, 67, 83, 113, 117, 119, 124, 125, 126, 127, 143, 158, 163
conductivity, 3, 146
constructed wetlands, 25, 30, 50
consumption, 12, 160
contaminant, 18, 45, 129
contaminated sites, 117, 139
contaminated soil, 19, 24, 25, 26, 29, 30, 31, 36, 44, 50, 109, 123, 129, 132, 133, 136, 139, 140
contaminated soils, 19, 24, 25, 44, 50, 109, 129, 132, 136, 139
contaminated water, 39, 143
contamination, 2, 22, 32, 35, 45, 48, 56, 65, 101, 104, 105, 106, 107, 110, 113, 114, 115, 128, 144, 153, 170, 178, 182
controversial, 126
cooking, 144
cooperation, 160
coordination, 58, 117
copolymers, 173
copper, 5, 15, 18, 19, 21, 23, 27, 29, 31, 36, 38, 42, 46, 49, 50, 65, 74, 101, 102, 103, 106, 108, 109, 110, 114, 129, 133, 135, 137, 139, 152, 153, 154, 162, 165, 168, 169, 175, 178, 179, 181
corrosion, 30, 48, 144
cosmetics, 144
cost, x, 33, 40, 63, 101, 102, 107, 110, 115, 127, 131, 142, 145, 155, 156, 157, 158, 160, 166
Cr, 2, 3, 5, 12, 16, 18, 21, 28, 30, 33, 37, 38, 39, 41, 43, 48, 50, 84, 85, 88, 89, 90, 91, 92, 93, 94, 98, 99, 100, 101, 104, 105, 107, 108, 114, 116, 121, 122, 127, 134, 136, 142, 148, 149, 150, 151, 158, 159, 160, 165, 174, 176
crops, 31, 134
crude oil, 8, 14
cycles, 3, 142
cycling, 24, 28
cysteine, 118, 168

D

deficiency, 45, 117, 122
degradation, 23, 27, 33, 38, 41, 83, 114, 115, 125, 126
denitrification, 6, 12, 45
deposition, 78, 107
deposits, 5, 88, 95, 96, 97, 99
derivatives, 53, 54, 76, 80, 104, 110, 125, 126
detection, 57, 72, 142, 147, 148, 149, 151, 153, 154, 156, 157, 158, 161, 162, 163, 164, 165, 168, 169, 170, 171, 172, 173, 180
detoxification, 25, 29, 119, 122
developing countries, 99, 164
diabetes, 117, 143, 160
dietary intake, 130, 173
diseases, 54, 89, 117, 144, 160
disinfection, 16, 43, 58
distribution, 26, 49, 58, 116, 118, 131, 133
diversity, 14, 27, 93
DNA, 58, 62, 75, 116
double helix, 116
drainage, 18, 22, 24, 111
drinking water, 16, 24, 35, 45, 89, 96, 105, 108, 109, 143, 144, 145, 146, 151, 154, 155, 160, 161, 164, 167, 168, 169, 171, 172, 173, 178
drought, 83, 89, 122, 124
drug delivery, 14, 59, 60, 68, 100, 128, 160
drugs, 6, 60, 125, 144
dyes, 3, 13, 54, 59, 110

E

E. coli, 14, 18, 59
ecosystem, 5, 18, 114, 115
effluent, 6, 16, 24, 37, 91, 108
effluents, 23, 41, 44
electrochemical methods, 5, 142, 156, 160
electrode surface, 156, 157
electrodes, 149, 157, 161, 169, 177
electron, 37, 42, 56, 59, 60, 63, 85
electronic structure, 62, 66
electrophoresis, 74, 164
electroplating, 100, 145
emission, 114, 147, 148, 150, 151, 162, 181, 182
energy, 9, 47, 54, 56, 60, 78, 154
energy transfer, 54, 78
engineering, 29, 45, 56, 105, 130
environment, 1, 16, 18, 19, 25, 29, 31, 32, 33, 35, 36, 44, 46, 48, 50, 53, 54, 77, 80, 87, 88, 92, 93, 94, 96, 99, 101, 109, 113, 114, 119, 123, 126, 127, 128, 131, 136, 138, 141, 142, 145, 146, 154, 160, 173, 176, 178, 179, 181
environments, 19, 33, 88, 101, 123, 126, 141
enzyme, 20, 33, 116
enzymes, 118, 122, 125, 126, 131
EPA, 146, 159, 180
equilibrium, 19, 103, 105, 106, 108, 109
equipment, 127, 142, 144, 155, 158
eukaryotic, 8, 63, 79
European Union, 109, 159
evidence, 17, 40, 46
exclusion, 49, 159
exposure, 20, 28, 54, 88, 115, 116, 118, 120, 129, 133, 136, 138, 143, 160, 161, 170, 172, 174, 176
extraction, 37, 44, 153, 155, 156, 165, 166, 170, 174, 175, 176, 177, 179, 181

F

Fabrication, 67, 69, 70, 72, 82, 84
families, 119, 120
farmland, 130, 133
fertilizers, 3, 89, 114
fiber, 89, 149, 156, 170, 171
fibers, 6, 32, 48, 89
filtration, 99, 145, 160, 175, 178
flame, 147, 151, 152, 154, 155, 162, 164, 165, 166, 175, 177
fluid, 94, 97, 98
fluorescence, 56, 64, 81, 147, 148, 149, 159, 163, 169, 171, 175, 178
food, 19, 57, 108, 109, 115, 124, 142, 143, 144, 145, 153, 164, 165, 169, 178, 179, 182
food chain, 124, 142, 145
formation, 2, 24, 48, 144, 154
fractures, 143
France, 64, 144, 167
freshwater, 91
fruits, 94, 172
FTIR, 77
fuel cell, 24, 43, 47
functionalization, 78, 85
fungi, viii, 3, 6, 18, 26, 32, 34, 88, 95, 106
fungus, 47

G

galvanizing industry, 3
genes, 76, 118, 120, 122, 126, 135
genetic diversity, 48
genetic engineering, 122, 131
genetics, 143
genus, 89
Germany, 76, 182
germination, 27
glutathione, 118, 132
graphite, 47, 147, 152, 153, 172, 173, 178

Index

grass, 34, 48
groundwater, 35, 36, 37, 43, 44, 47, 48, 49, 109, 141, 178
growth, 7, 14, 21, 27, 34, 37, 43, 58, 117, 120, 122, 123, 124, 132
growth rate, 120, 122

H

habitat, 96, 124
habitats, 95, 114
hair, 170, 178
harmful effects, 142, 145
hazardous substance, 29, 45
hazardous substances, 29, 45
health, 13, 65, 78, 87, 99, 104, 108, 136, 143, 145, 161, 170, 176
health effects, 104, 170
heavy metal removal, 1, 4, 5, 19, 20, 26, 36, 53, 54, 56, 57, 87, 88, 92, 96, 101, 103, 105, 106, 107, 108, 113, 115, 169, 173, 176, 178, 179
heavy metal toxicity, 109, 139, 142, 143, 171, 179
heavy metals analysis, 142, 147, 158, 159, 171
heavy metals toxicity, 109, 139, 142, 143, 171, 179
Hg, 2, 13, 22, 23, 27, 34, 88, 92, 95, 98, 114, 116, 117, 121, 126, 134, 138, 142, 144, 148, 149, 150, 151, 154, 156, 158, 159, 160, 171, 173, 175, 176
histidine, 118, 123
human, 1, 15, 17, 20, 53, 60, 68, 74, 87, 88, 115, 116, 126, 132, 135, 142, 144, 145, 146, 170, 175, 178, 181
human body, 53, 143, 144
human health, 88, 115, 126, 142, 145, 146, 178, 181
hydrocarbons, 35, 135
hydrogen, 13, 25, 144
hydroxide, 14, 95, 97, 166
hyperaccumulators, 114, 119, 120, 122, 123, 128, 129, 130, 131, 132, 133, 134, 135, 136, 137, 138, 139, 140

I

ideal, 6, 90, 95
identification, 173, 175
immobilization, 17, 31, 33, 42, 103
immune system, 134, 144
impurities, 2, 90, 115, 117
in vitro, 16, 44, 135
India, 23, 89, 107, 171
induction, 69, 116, 154
industries, 2, 145
industry, 2, 53, 65, 114, 144, 176
infertility, 117, 160
ingestion, 143, 145
inhibition, 27, 117
interface, 76, 78, 157, 159
interference, 153, 155
ions, 6, 25, 27, 36, 39, 42, 49, 50, 88, 91, 92, 93, 94, 95, 96, 97, 98, 99, 100, 101, 104, 105, 108, 109, 110, 116, 119, 129, 142, 143, 144, 147, 154, 156, 157, 158, 159, 165, 166, 168, 169, 175
iridium, 58, 152, 157
iron, 6, 7, 12, 14, 15, 17, 20, 24, 27, 29, 33, 34, 36, 40, 41, 42, 43, 44, 45, 46, 48, 50, 54, 65, 68, 105, 149, 152, 153, 165, 176, 177, 178, 181
irrigation, 96
IRT, 120
Islam, 127, 132
isotherms, 106, 108
issues, 104, 143

K

kidneys, 89, 102, 116, 129, 134, 143, 160

kinetic model, 20, 102
kinetics, 12, 58, 103, 104, 106, 108, 132, 165

L

lactoferrin, 15, 34
lakes, 141, 165
lanthanide, 81, 142
lead, 3, 6, 17, 18, 20, 23, 26, 27, 30, 33, 41, 46, 54, 101, 103, 104, 108, 109, 114, 116, 129, 133, 134, 136, 138, 140, 142, 144, 152, 153, 154, 161, 162, 164, 168, 174, 175, 176, 177, 178, 181, 182
light, 54, 57, 59, 73, 147, 152, 158
lignin, 3, 90, 104, 106, 110
liquid chromatography, 142, 150, 158, 177, 179
liquid phase, 170, 182
liver, 54, 89, 116, 129, 134, 143, 144, 160, 170, 174
locus, 116, 135
Luo, 45, 57, 67, 72, 76, 82, 133, 180

M

magnitude, 13, 119
management, 22, 135, 177, 182
manganese, 18, 24, 36, 40, 48, 96, 135, 139, 152, 153, 154, 174, 178
mass, 49, 58, 142, 147, 148, 150, 153, 158, 172, 173, 176, 177, 179
mass spectrometry, 49, 58, 142, 147, 148, 150, 153, 158, 172, 173, 176, 177, 179
materials, 5, 30, 53, 54, 56, 71, 78, 79, 87, 97, 99, 145, 153, 157, 164, 165
matrix, 40, 152, 181
measurement, 156, 171
measurements, 62, 63, 161
medicine, 54, 60
membranes, 3, 39, 58, 68, 105, 175

mercury, 2, 12, 13, 20, 21, 45, 47, 54, 101, 106, 114, 128, 134, 137, 138, 139, 142, 157, 163, 164, 165, 168, 172, 175, 176
Mercury, 2, 14, 22, 23, 28, 49, 146, 163, 176
metabolism, 3, 41, 122, 141, 143
metal ion, 2, 28, 40, 87, 88, 92, 94, 95, 96, 98, 101, 102, 104, 106, 107, 109, 110, 119, 120, 127, 141, 142, 147, 152, 156, 158, 160, 162, 165, 170, 171, 173, 175, 182
metal ions, viii, ix, 2, 28, 87, 88, 92, 94, 95, 96, 98, 101, 102, 104, 106, 107, 109, 110, 119, 127, 141, 142, 147, 152, 156, 158, 160, 162, 165, 170, 171, 173, 175, 182
metal recovery, 5, 47
methodology, 28, 32, 162
methylene blue, 72, 83
mice, 62, 66
microbial activity, 1, 13, 21, 45, 126
microbial communities, 18, 24, 27, 131
microbial community, 30, 32, 37, 45, 46, 50
microcosms, 17, 28
microorganisms, 1, 2, 4, 5, 6, 7, 8, 12, 14, 17, 18, 19, 20, 22, 23, 26, 28, 29, 31, 35, 36, 37, 38, 39, 40, 43, 44, 47, 48, 49, 51, 53, 54, 105, 125
mineralization, 32, 47
Miscanthus, 123, 132
modifications, 88, 92, 93, 94, 97, 99, 100
molecules, 59, 68, 118, 120
molybdenum, 31, 153, 154
MRI, 14, 59, 100, 128, 160

N

nanocomposites, 5, 56, 64, 83, 108
nanomaterials, 44, 60, 68, 71, 157
nanoparticles, 5, 12, 14, 25, 32, 38, 41, 49, 51, 57, 68, 69, 149, 171

Index

nanosystems, 57, 68
nanotechnology, 57, 84, 99
natural habitats, 94, 95, 97
neutrophils, 48, 144
nickel, 2, 15, 19, 21, 27, 36, 39, 43, 102, 106, 108, 109, 110, 114, 132, 138, 144, 152, 153, 154, 161, 166, 167, 175, 176, 178, 179, 181
nitrogen, 3, 25, 26, 46, 48, 50, 54
nutrient, 14, 122

O

oceans, 53, 141
oil, 2, 26, 48, 106, 154, 175
optimization, 13, 162, 169, 170, 179
organelles, 63, 119
organic compounds, 12, 33, 54, 90, 114, 135
organic matter, 16, 37, 125, 155, 178
organism, 27, 58, 61, 115, 126
organs, 119, 143, 144
oxidation, 14, 16, 17, 20, 23, 31, 33, 35, 38, 41, 56, 117, 118, 143, 146, 156
oxidative stress, 13, 134, 144, 167, 174
oxide nanoparticles, 25, 45, 49, 149, 165
oxygen, 2, 6, 12, 50, 77, 116, 137, 144
oxygen deficits, 2
oyster, 95, 99

P

Pakistan, 162, 163
palladium, 32, 152
pathogenesis, 137, 180
pathway, 27, 45
pathways, 113, 143, 145, 169
Pb, 2, 4, 13, 37, 45, 48, 61, 83, 88, 89, 90, 91, 92, 93, 94, 95, 96, 98, 100, 102, 103, 105, 107, 108, 111, 114, 116, 117, 121, 122, 123, 125, 127, 130, 132, 134, 138, 142, 144, 148, 149, 150, 151, 154, 158, 159, 160, 165, 166, 173, 175, 176, 177
PCP, 4, 29, 47, 77
peptides, 9, 42, 119, 149, 164, 177
petroleum, 30, 33, 59, 125, 135
pH, 3, 12, 21, 37, 67, 93, 94, 98, 117, 122, 125, 146
pharmaceutical, 18, 32
pharmacokinetics, 59, 62
phenol, 4, 37, 47, 90
phosphate, 31, 41, 49, 93, 96, 97, 103
phosphorescence, 64, 78
phosphorus, 21, 23, 26, 32, 46, 47, 48
photoactive materials, 53, 54, 74
photobiology, 55
photochemistry, 55, 57
photodynamic therapy, 54, 84
photosynthesis, 2, 56, 83
physiology, 118, 119
phytodegradation, 114, 115, 125, 131, 135, 139
phytoextraction, 31, 114, 115, 122, 123, 129, 130, 131, 132, 134, 135, 136, 138
phytofiltration, 36, 114, 115, 126, 132, 135, 139
phytoremediation, 24, 26, 38, 48, 50, 113, 114, 115, 117, 122, 124, 126, 127, 128, 129, 130, 131, 132, 133, 135, 136, 137, 138, 139, 140
phytovolatilization, 114, 115, 126, 133
plant growth, 41, 118
plants, 2, 3, 5, 18, 53, 59, 87, 88, 89, 106, 113, 115, 117, 118, 119, 120, 122, 123, 124, 125, 126, 127, 128, 130, 131, 132, 134, 135, 136, 137, 138
platform, 80, 168, 171
platinum, 32, 67, 152, 157
PM, 21, 34, 38, 44
Poland, 1, 87, 113, 141
pollutants, 2, 5, 13, 18, 38, 59, 64, 87, 88, 103, 107, 115, 116, 125, 126, 127, 165

pollution, 3, 5, 8, 9, 24, 53, 54, 88, 99, 114, 125, 130, 132, 135, 139, 140, 146, 164
polymer, 38, 58, 79, 90, 110, 171, 181
polymer nanocomposites, 79, 110
polymers, 54, 56, 107, 157
precipitation, 5, 12, 42, 43, 49, 124, 145, 155, 160
preparation, 79, 155, 156
prevention, 32, 119, 178
produce reactive oxygen, 54
proliferation, 117, 118
protection, 114, 119
proteins, 6, 22, 25, 27, 54, 57, 65, 71, 116, 118, 120, 122
Pseudomonas aeruginosa, 7, 27, 75
purification, 25, 54, 58, 89, 93, 114, 115, 126, 127, 128, 145
PVC, 68, 137
pyrolysis, 94, 172

R

radiation, 21, 147
Radiation, 41, 171
raw materials, 99, 100
reaction center, 58, 62, 63, 66, 82
reactions, 14, 56, 116, 118, 119
reactive oxygen, 54, 116, 137, 144
reactivity, 38, 43, 55
recovery, 28, 48, 50, 83, 100
recycling, 88, 110
red mud, 146, 161
regulations, 146, 180
remediation, 3, 7, 8, 9, 13, 17, 18, 22, 24, 26, 35, 45, 48, 96, 97, 99, 109, 114, 115, 123, 127, 131, 132, 133, 139
remnants, 3
repair, 60, 62
researchers, 3, 97, 146
residues, 18, 60, 89, 92, 95, 98, 102, 106, 161

resistance, 14, 23, 26, 30, 35, 56, 117, 135, 175
response, 2, 25, 28, 31, 120, 123, 134, 135, 138
ribonucleotide reductase, 20, 59
rice husk, 93, 100, 101, 104, 105
rights, iv
risk, 13, 23, 130, 138, 139, 143, 145, 157, 161, 166, 167, 172, 180
risk assessment, 139, 161, 172, 180
risks, 139, 170
root, 6, 16, 41, 119, 122, 124, 127, 128
root system, 122, 124
roots, 92, 113, 117, 119, 124, 126, 127
routes, 143, 160

S

salinity, 35, 47
Salmonella, 14, 42
salts, 59, 175
sawdust, 90, 102, 105, 108, 110
sediment, 35, 36, 40, 45, 103, 125, 132
sediments, 17, 39, 44, 45, 161, 168
seed, 27, 108, 146, 166
seedlings, 45, 126, 139
selectivity, 6, 45, 73
selenium, 46, 126, 153, 154, 173, 174, 175, 180
self-cleaning process, 2
semiconductor, 56, 65
sensing, 72, 162, 165, 168, 171
sensitivity, 57, 142, 153, 154, 156
sensor, 171, 180
sensors, 54, 57, 164
sequencing, 4, 15, 29, 36, 45, 46
sewage, 2, 3, 31, 89, 105, 114
shoot, 6, 117
shoots, 119, 122, 126
signal transduction, 118, 143
signals, 85, 156

Index

silica, 27, 53
silver, 12, 14, 16, 17, 22, 25, 30, 32, 41, 43, 48, 51, 75, 149, 152, 153, 154, 157, 168, 171
silver chloride nanoparticles, 12
skin, 13, 77, 89, 94, 143, 160
slag, 34, 39
sludge, 2, 5, 13, 14, 18, 19, 24, 25, 26, 29, 31, 36, 37, 41, 42, 43, 45, 46, 47, 48, 49
sodium, 31, 95, 96, 152, 154
soil pollution, 114, 124
solid phase, 37, 159, 165, 177, 181
solidification, 156, 162, 168
solubility, 13, 59, 95, 124, 144, 159
solution, 3, 5, 19, 40, 49, 51, 91, 92, 94, 98, 100, 101, 102, 104, 105, 108, 109, 110, 111, 139, 152, 157, 158, 166, 167, 175, 176
sorption, 3, 5, 12, 88, 91, 92, 101, 104, 165
speciation, 129, 158, 161, 162, 164, 165, 166, 168, 170, 173, 174, 179
species, 5, 12, 39, 49, 53, 54, 89, 90, 91, 93, 95, 96, 98, 103, 113, 116, 117, 118, 119, 120, 121, 123, 125, 128, 130, 131, 132, 135, 137, 138, 139, 144, 156, 163, 169, 175, 177, 179
spectroscopy, 12, 149, 151, 155, 159, 165, 169, 180, 181, 182
stabilization, 114, 125
state, 2, 13, 24, 68, 117, 126, 130, 152
states, 13, 143, 158
stomata, 125, 126
storage, 14, 26, 118, 122, 123, 124
stratification, 2, 30
stress, 5, 27, 29, 38, 131, 135
structure, 32, 36, 37, 38, 46, 63, 66, 79, 90, 96, 99, 116, 128, 134
sulfate, 4, 12, 17, 18, 20, 23, 39, 161
sulfur, 3, 13, 23, 58
Sun, 8, 18, 21, 30, 34, 42, 47, 48, 50, 57, 61, 70, 74, 84, 108, 131, 140, 159, 166, 179
supplementation, 20, 76
surface area, 93, 97
surfactant, 20, 45, 104, 175
surfactants, 2, 23, 24, 97
synthesis, 27, 29, 54, 58, 68, 80, 83, 117

T

Taiwan, 89, 100, 109, 110
target, 54, 155
techniques, 6, 8, 10, 12, 44, 57, 69, 87, 92, 105, 127, 142, 147, 155, 156, 157, 159, 160, 162, 164, 167, 171, 177
technologies, ix, 5, 8, 22, 25, 56, 57, 99, 113, 132, 145
technology, 12, 28, 55, 103, 113, 115, 124, 127, 131, 137, 145, 173, 179
temperature, 3, 12, 21, 64, 93, 154, 163
thallium, 152, 154, 172
therapy, 66, 167
thermodynamics, 106, 108
tin, 58, 152, 153
tissue, 57, 115, 116, 118, 136, 142
titanium, 72, 84
tobacco, 98, 107
toxic metals, 116, 117, 122, 127, 135, 136
toxicity, 6, 22, 23, 34, 49, 61, 102, 115, 118, 128, 139, 140, 142, 143, 144, 163, 171, 179
toxin, 57
trace elements, 32, 135, 136, 153, 161, 180
transformation, 18, 33, 77, 116, 124, 125
translocation, 6, 118, 119, 122
transport, 3, 53, 59, 88, 95, 120, 127, 135, 138, 164
transportation, 2
treatment, 2, 5, 12, 15, 16, 22, 23, 24, 25, 26, 28, 31, 32, 36, 38, 39, 40, 41, 42, 44, 46, 47, 55, 68, 96, 97, 105, 108, 109, 116, 144, 163, 171
trial, 17, 32

U

ultrasound, 97, 177
uranium, 21, 22, 24, 26, 28, 31, 32, 35, 36, 37, 40, 43, 46, 49, 154
urban, 15, 88, 101, 107, 114, 164
urbanization, 87, 141, 142

V

vaccine, 144
Valencia, 12, 37
vanadium, 27, 29, 37, 47, 49, 130, 154
vapor, 144, 147, 148, 152, 154
vegetables, 164, 166, 172
vegetation, 117, 118

W

waste, 2, 3, 5, 19, 24, 26, 43, 48, 54, 61, 64, 88, 90, 92, 94, 97, 98, 99, 101, 104, 106, 107, 110, 127, 145, 155, 163, 169, 176
waste water, 26, 43, 127, 176
wastewater, 2, 5, 12, 15, 16, 17, 19, 20, 23, 25, 26, 27, 28, 29, 31, 32, 34, 35, 36, 38, 39, 41, 42, 45, 46, 47, 50, 54, 59, 68, 84, 88, 89, 91, 98, 100, 101, 102, 103, 105, 106, 107, 108, 123, 128, 138, 144, 145, 154, 157, 160, 161, 162, 165, 169, 171, 176, 179
water purification, 53, 163
water quality, 13, 180
water vapor, 93, 97
wetlands, 134, 139
WHO, 130, 146, 151, 159
wood, 89, 90
workplace, 65, 160
World Health Organization, 146, 151, 159, 180

Y

yeast, 29, 58, 135
yeasts, 3, 29, 40, 44, 58, 75, 88, 89, 135

Z

zeolites, 55, 97, 104
zinc, 15, 29, 33, 37, 44, 49, 62, 79, 106, 107, 110, 114, 134, 138, 139, 140, 142, 152, 154, 164, 175, 178, 181